Essential Math for Physics Undergraduates

Essential Math for Physics Undergraduates: Practical Examples with Python Tools covers basic applied Math and entry-level programming. A gap often exists in the Physics student's Math curriculum, between the Math taught by the Math departments and the Math taught either in a "Math Methods" or advanced Physics courses, such as introduction to quantum mechanics, classical mechanics, thermal and statistical physics, and electromagnetic theory. This text plugs that gap and provides the new Physics student with the necessary tools to succeed in their Physics curriculum. The book helps instructors to fill in the gap between what new Physics students learn in high school and introductory college math courses, providing them with what they need to know as they progress through their Physics curriculum.

Key Features:

- Covers Math that is essential in Physics curricula.
- Offers a more accessible introduction to Math than can be found in typical "Math Methods for Physicists" texts.
- Introduces Python and shows the reader how to use it for basic math problems.

Brett DePaola is a professor of Physics at Kansas State University. He is an experimentalist in the field of atomic, molecular, and optical physics. Born in Ohio, he received his BS in Physics at Miami University, and his MS in Physics at Miami University, for research done at The University of Paris in Orsay, France. He then received his PhD in Physics from The University of Texas, Dallas. For many years DePaola has taught courses designed to help first-year Physics students prepare themselves for their undergraduate Physics curriculum..

Essential Math for Physics Undergraduates
Practical Examples with Python Tools

Brett DePaola

CRC Press
Taylor & Francis Group
Boca Raton London New York

CRC Press is an imprint of the
Taylor & Francis Group, an **informa** business

Designed cover image: Shutterstock

First edition published 2026
by CRC Press
2385 NW Executive Center Drive, Suite 320, Boca Raton FL 33431

and by CRC Press
4 Park Square, Milton Park, Abingdon, Oxon, OX14 4RN

CRC Press is an imprint of Taylor & Francis Group, LLC

© 2026 Brett DePaola

ISBN: 978-1-032-97098-1 (hbk)
ISBN: 978-1-032-98376-9 (pbk)
ISBN: 978-1-003-59828-2 (ebk)

DOI: 10.1201/9781003598282

Typeset in LM Roman
by KnowledgeWorks Global Ltd.

Publisher's note: This book has been prepared from camera-ready copy provided by the authors.

I dedicate this book to Professor George B. Arfken, an outstanding teacher and a wonderful person.

Contents

Preface

There are a few really good texts available for the expressed purpose of teaching a physics student "all the math they need to know" at both the advanced undergraduate and graduate level. For algebra, trigonometry, and elementary differential and integral calculus, physics departments usually rely on their university's math department. But we have observed that a gap often exists in the physics student's math curriculum, between the math taught by the math departments and the math taught either in a "math methods" course or as part of the advanced physics courses, such as introduction to quantum mechanics, classical mechanics, thermal and statistical physics, and electromagnetic theory. The purpose of this text is to plug that gap.

Of necessity, some of the material will overlap what is taught to physics undergraduates by some math departments; other material in this text will overlap what is taught in a math methods course. If so, the instructor can skip that material as they work through this text.

An introduction to computer programming is also considered essential in the modern physics curriculum. As in many other physics departments, we have hit upon Python as a suitable language for our students. Because it is outside the range of this text to provide a comprehensive tutorial in Python, we have included only a brief primer to summarize the language. This allows us, however, to include brief Python scripts in the other chapters, in order for the student to see how one could use them to simplify otherwise tedious tasks, for example, matrix manipulations. If the instructor so chooses, these Python sections, as well as the Python primer itself, can be omitted.

Brett DePaola
Department of Physics
Kansas State University
January 2025

## CHAPTER 1

A Python Primer

1.1 THE BASICS OF THE LANGUAGE

The Python programming language is rapidly becoming one of the most popular computer languages, particularly in artificial intelligence and in scientific computing. It is a so-called *interpreted language*, meaning that the user need not *compile* the code (convert into machine language) before executing it. Unlike many other languages, such as C, in Python one does not specify the type of variable. That is, one does not directly indicate if a variable should be treated as an integer, floating point, string, *etc.* Some may like this feature; others may dislike it. But all agree that not being a *strongly typed* language makes for less typing in Python than in other languages.

Python is freely available. One of the easiest ways to install it is by downloading and installing the Anaconda package. (See Appendix A.) This free package can be installed on all major computer platforms and comes complete with all of the most used add-on packages. Some of these packages, `scipy`, `numpy`, and `matplotlib`, will be described in this chapter.

There are many outstanding texts on Python. One can also find online many fine Python tutorials, many of them free. We include this brief primer on Python for the sake of completeness, since in subsequent chapters we will include sample Python code, designed to aid the student in treating the various math topics numerically as well as analytically.

1.2 VARIABLES

1.2.1 Variable Names

You can use nearly any combination of characters for your variable names, but there are a few restrictions.

- A variable name must start with a letter or an underscore (`_`), but cannot start with a number. Thus, `abc5` is valid, but `5abc` is not.
- A variable name cannot be a Python keyword such as `if`, `else`, `for`, `while`, *etc.*
- A variable name cannot contain a space. Thus, `abc` is a valid variable name, but `a bc` is not.

Python is *case sensitive*. That is the variable `Gorp` is not the same as the variable `gorp`. It is a useful practice to use variable names that indicate what that variable means. For example, if you are using a variable to contain the value of a velocity, you probably should not use `x`, even though Python would not object. A better choice might be `v` or `vel`. If you'd like even more information contained in your variable name, you may use multiple word names – as long as no spaces are contained within the name. Multi-word names generally follow one of two formats, the so-called "camelcase" and the so-called "underscore". An example of the former is `velocityRock`; an example of the latter is `velocity_rock`. Note that both follow the above rules for naming variables, and both are made easier to read by separating the words, either using capital letters or underscores. I suggest that you choose one or the other convention and stick to it. Python won't care if you mix conventions, but you may confuse yourself when you try to reference a variable later in your code and use the wrong convention!

There are many variable types in Python. The most commonly used of these are *integers*, *floating points*, *strings*, *lists*, *tuples* and *boolean*.

1.2.2 Integers

Integers are non-floating point numbers. The range of integer values in Python-3 is limited only by the available memory on your computer, and include both positive and negative values. An example of a single line of code in which an integer variable is defined is

```
iteration_number = 27
```

Note that the product of two *integers* is, itself, an *integer*.

1.2.3 Floating Points

We can similarly define a floating point variable:

```
acceleration = 9.8
```

Within the computer, floating point numbers are represented with 64 bit values. The magnitude of the largest allowed floating point number is about 1.8×10^{308}. The magnitude of the smallest allowed floating point number is about 2.2^{-308}. You can limit the number of decimal places in a floating point number using the **round** method. (A *method* is a sort of built-in function associated with a *class* of variable.) An example using the **round** method is

```
round(3.14159,2)
```

The result of this operation would be `3.14`. The product (or sum) of a *float* with either another *float* or with an *integer* is a *float*.

1.2.4 Strings

Even in scientific computing, it is often necessary to work with characters. For example, if you want to write your computed numbers to a file, you must specify the name of the file. (File input/output will be discussed in Section 1.4.2 of this chapter.) To facilitate working with characters, Python has a variable type called a *string*. Strings are very simple to use and manipulate thanks to the many *methods* that are part of the Python language. Here's a simple example:

```
1  '''
2  Examples of Using Strings
3  Written 10/11/2025 by B. DePaola
4  '''
5  s1 = 'Big'
6  s2 = "Kahunas "
7  print(s1)
8  print(s2)
9  s3 = s1 + ' ' + s2   # Note the inserted space
10 print(s3)
11 print(5*s3)
12 print(str(3) + ' ' + s3)
```

Listing 1.1: Sample code creating and manipulating strings.

When executed, Listing 1.1 will give the following output:
```
Big
Kahunas
Big Kahunas
Big Kahunas Big Kahunas Big Kahunas Big Kahunas Big Kahunas
3 Big Kahunas
```

There are several things to note about this, our first Python program:

First, observe lines 1-4. This is a *comment block*. A comment block begins and ends with three single quotes. Everything that lies between a pair of triple quotes is a comment and will not be interpreted as code to be executed. You should always start your programs with a comment block that, at the least, tells what the purpose of the code is, as well as when and by whom it was written. A second way to place comments in your code is the use of the "hash" (#), as shown in line 9. Python treats everything following a hash as a comment, up to the end of the line. Some people prefer to create a comment block using only hashes. In this case, every line in the comment block must begin with a hash.

Next look at lines 5 and 6. Here I defined two strings, s1 and s2. I could have given these strings any name I want, but these names seemed appropriate and simple. Python knows that s1 and s2 are strings because the quantities on the right of the equal sign are enclosed in quotes. Note that you can use either single or double quotes, but whichever you use on the left side of the "phrase" you must use on the right side as well. Most people use single quotes, but there is an advantage to using double quotes. For example, suppose I want to define the string monkey's uncle. Using single quotes around this would not work because the single quote in the string would mess things up. Here I would need to surround the phrase with double quotes.

Now look at lines 7, 8, and 10-12. These **print** statements send the value of whatever is contained in the following parentheses to the screen. (We'll deal more with **print** statements in Section 1.4.1.)

Next examine line 9. This demonstrates how to combine strings. Here I have combined s1 and s2 and stored that concatenation in a third string, s3. Line 11 shows what happens when you multiply a string by an integer: you get that same string repeated, in this case five times.

Unlike in some programming languages, Python does not automatically convert numbers into strings. In line 12, in order to tell Python that I want "3" to be treated as a string and not a number, I use the `str` method.

Concatenating and repeating strings is great, but what if you want to pick out a particular character in a string. There are several ways to do this. One is with a process known as *slicing*. Here are some examples, where it is assumed that s1, s2 and s3 are defined as above.

```
1  print(s3)
2  print(s3[0])
3  print(s3[4])
4  print(s3[2:8])
5  print(s3[:9])
6  print(s3[4:])
7  print(s3[0:-1])
8  print(s3[4:-4])
```

Listing 1.2: Sample code showing string slicing.

The output from this code will be:
```
Big Kahunas
B
K
g Kahu
Big Kahun
Kahunas
Big Kahunas
Kahu
```

You can think of a string as a sort of array of characters. (Note that "array" actually refers to a specific data type that we will discuss later in this chapter. Here I am using the term in a non-data-type sense.) If you wish to look at a specific element in a string, you write down the string name, followed by a pair of square brackets. Within the square brackets you indicate the location of the string that you wish to print or use. In line 2, we look at the very first character in the string, s3. Note that a string's indices start at 0, not 1. In line 3, we print the character in the 4th position (starting from 0!) in the string. But as line 4 shows, you can also access a range of characters in a string. The 2:8 means to begin at the character in position 2 and end with the character just before the character in position 8. As lines 5 and 6 show, leaving out the starting or ending index implies that you want to start from the beginning or finish with the end character.

What if you don't know how long the string is, but you'd like to finish with the end character? Well, there is a method for finding the length of a string. In the case of our example string, you would use `len(s3)`. But there is a simpler way. Thus far, we start with beginning character being at position 0, and each subsequent character's position incremented by +1. But instead of moving character by character to the right, you could imagine that the string wraps around on itself. Now counting by -1 moves you to the left of 0 by one character. In other words, to the final position of the string. In line 7 we print out the entire string, from position 0 to position -1. In line 8 we print the character in position 4 (always starting from 0) through the 4th character from the right.

Method Name	Purpose
s.lower()	returns the lower case of string **s**
s.upper()	returns the upper case of string **s**
s.strip()	returns string **s** but with all white spaces stripped from the start and end
s.replace('old', 'new')	returns **s**, but with all instances of 'old' replaced by 'new'
s.split('delimiter')	returns a list of substrings from **s**, split by the given delimiter

These are the essentials of string manipulation, but there are many string methods. Here is a table of the most commonly used ones:

1.2.5 Lists

In some ways, *lists* are much like *strings*, but there are also major differences. I cannot stress enough how important *lists* are. So what is a *list*? Like a string, a list is an "array" of elements. But unlike a string, elements of a list can be of any variable type. One way to create a list is to put the elements of list within a pair of square brackets, with each element separated with commas. Let's illustrate by some examples.

```
1  '''
2  Examples of creating and using lists
3  '''
4  mylist1 = [27, "oak tree", 3.14159]
5  print(mylist1)
6
7  mylist2 = []
8  print(mylist2)
9  mylist2.append("new element")
10 print(mylist2)
```

Listing 1.3: Sample code showing how to create lists.

```
[27, 'oak tree', 3.14159]
[]
['new element']
```

Line 4 of the above code shows the classic list creation. Notice first that in the creation statement, the elements of the list are placed within square brackets. Next note that the different elements in a list need not be of the same data type. In this example, the first element is an integer, the second is a string, and the third is a float. In line 7 I created an "empty list". This is very often a useful construct. In line 9 I then proceeded to add an element to that empty list. When we get to loops in Section 1.3, we will see how useful the **append** method is.

All the slicing techniques we learned for strings apply equally as well to lists. Thus, we can isolate any element(s) of a list that we wish. We'll see more about how to use lists later in this chapter.

1.2.6 Tuples

A tuple is almost exactly like a list. There are only two differences: First, when creating a tuple, one uses curly brackets instead of the square brackets we employ for creating lists. Second, and most importantly, tuples are *immutable*. That is to say, while I can replace any element of a string or list with some other value, I cannot do this with a tuple. But why do I need both tuples and lists? Well, suppose I have an "array" of elements and I do not wish to accidently change any of the elements of that array? In this case, it is preferable to use a tuple rather than a list. As we will see, many of the arguments to the various methods in Python are given in the form of a tuple. Because of their close similarity to lists, we need say no more about tuples.

1.2.7 Booleans

Booleans are trivial. They are basically just the lowest level of logic elements. A boolean has a value of **True** or **False**. You can assign one of these values to a variable of type boolean, or you can apply a boolean operator and make a decision on what that result is (True or False). We'll see lots of implicit uses of booleans when we discuss flow control later in this chapter.

```
1  '''
2  Examples of using Booleans
3  '''
4  a = True
5  b = False
6  print(a, b)
7  print(1 == 1)
8  print(1 == 2)
9  print(4 > 5)
```

Listing 1.4: Sample code showing how to use booleans.

```
True False
True
False
False
```

In lines 4 and 5 of the above code, we make a direct assignment of variable to the key words **True** and **False**. Note the required use of upper case on those key words. A more common way of using booleans is through the use of conditional operators, as shown in lines 7-9. Earlier we learned that the equal sign is used to assign values to variables. In lines 7 and 8 we use a "double equal sign" to indicate a boolean operator. Boolean operators can be thought of as making a statement, and Python then replies by telling us if the statement is True or False. In the case of the double equal sign, the statement is, "the left side is equal to the right side. We see the example that this is a True statement in line 7, but is a False statement in line 8. In line 9, the statement is, "4 is greater than 5". This is, of course, a False statement. In the next section we will use boolean operators to help us with a program's flow control.

1.3 FLOW CONTROL

1.3.1 while-loops

Here is a small sample of a while-loop:

```
1  '''
2  Example of using a while-loop to control flow
3  '''
4  counter = 0   # initialize my counter variable
5  while (counter < 4):
6      print("I'm still counting. Counter = ", counter)
7      counter += 1   # increment the counter
8  print("All done counting. Counter = ", counter)
```

Listing 1.5: Sample code showing how to use a while-loop.

Resulting in the output:
```
I'm still counting. Counter = 0
I'm still counting. Counter = 1
I'm still counting. Counter = 2
I'm still counting. Counter = 3
All done counting. Counter = 4
```

The purpose of this code is to use a *loop* to repeat a step multiple times. In Python, you can readily recognize a *block* of code through indentation. *The indentation is required*, although how many spaces you indent is your choice. In this case, the block shows the extent of the loop. The loop starts with a **while** statement. Following the keyword, **while**, we have a boolean statement in parentheses, followed by a colon. In this example, the boolean statement tests whether or not the variable **counter** is less than 4. (By the way, do not forget to precede the loop by a statement in which you give your variable a value!) If the boolean statement is True, the loop will be executed. If the statement is False, the loop is skipped. In this case, the contents of the loop are trivial: just a **print** statement and the all-important modification of the loop variable. If I were to omit this step, **counter** would always be less than 4, and I would have what is called an "infinite loop". Line 7 is interesting. What it means is, "the new value of **counter** is equal to the old value plus one." This is a very common construct. As you can see, the loop is executed 4 times. But at the end of the 4th iteration, the variable finally reaches the value 4. Thus, when the program returns to the conditional, the statement is no longer True and the program jumps to the first statement that follows the loop.

1.3.2 for-loops

This listing shows another kind of loop, the *for-loop*.

```
1  '''
2  Example of using a for-loop to control flow.
3  '''
4  for i in range(5):
5      print(i)
6  print("All done!")
```

Listing 1.6: Sample code showing how to use a for-loop.

This results in the output:

```
0
1
2
3
4
All done!
```

The first thing to notice is that the general structure of a **for**-loop is very similar to that of a **while**-loop: They both start with a conditional followed by a colon, which is then followed by an indented block. The **range** function returns a sequence of numbers. The full syntax of **range** is

range(begin, end, step-size)

where **begin** is an optional integer telling **range** where to begin its sequence, **end** is an integer telling **range** where to stop short of. (The value of **end** is not included in the output.) The optional argument **step-size** gives the increment; the default is 1.

The **for**-loop makes use of two key words, "**for**" and "**in**". Operationally, what happens is that the **for-in** pair steps its way through all the values that follow, one by one executing the commands in the following block.

It is not an exaggeration to say that loops are the heart and soul of all programming languages.

1.4 INPUT/OUTPUT

1.4.1 Input from the Keyboard and Output to the Screen

We have already seen the simplest way to send an output to the screen via the **print** command. But we can do so with much greater sophistication using *formatted output*. The nice thing is that the same formatting commands that work for printing to the screen will work when printing to a file – which we will discuss in the next section. The most commonly used way to format outputs are:

– Using the String Modulo Operator (%)

– Using the Format Method

The String Modulo Operator is, in some way, the clunkiest approach, but its chief advantage is that it is virtually identical to that used in formatted outputs in the C language. Here are some examples:

```
1  '''
2  Examples of Modulo Operator Formatted Output
3  '''
4  my_variable = 'something'
5  text1 = 'some'
6  text2 = 'more'
7  print("I'm going to inject %s here." %my_variable)
8  print("I'm going to inject %s text here, and %s text here." \
9       %(text1,text2),'\n')
10
11 # Here are some more examples using numbers:
12 my_pi = 3.14159
13 print("unformatted:",my_pi)
14 print("Formatted with one output: %5.3f" % my_pi)
15 my_e = 2.71828                    # base of the natural logs...
16 print("Formatted with two outputs: %5.3f %4.5f" % (my_pi, my_e))
```

Listing 1.7: Sample code showing how to format output.

This results in the output:
```
I'm going to inject something here.
I'm going to inject some text here, and more text here.
unformatted: 3.14159
Formatted with one output: 3.142
Formatted with two outputs: 3.142 2.71828
```

As shown in line 7, the basic format of the *modulo operator* technique is that you insert a % followed by a letter where you want your variable's value to be placed, all of this inside a string. Different letters are used to indicate different variable types. In this case, the variable is a string. Then, later in the print statement, outside the string, the % appears again, this time followed by the variable whose value is to be printed. Here is a table indicating which letter to use for which variable type:

specifier	used for
c	character
d or i	signed decimal integer
e or E	scientific notation with e or E
f	decimal floating point
g or G	use the shorter of e, E, or f
s	string of characters
u	unsigned decimal integer

Line 8 shows how to print the values of more than one variable. Note that the all the variable values are placed in a tuple. The first element in the tuple goes where the first % is, *etc.*

Lines 14 and 16 shows how to format floating point values. The number before the decimal point indicate the total number of characters in the output (including the decimal point). The number following the decimal point indicates the number of decimal places in the output.

Modulo operator formatting is easy and, as already stated, is virtually identical to formatting in other languages, including C. But there are other, more modern, ways to format. However there is a newer "Pythonic" way to format output. If you are not already hooked on the formatting methods from other computer languages, I recommend that this is what you should learn and use.

The syntax is:

`'String here { } then also { }'.format('something1','something2')`

That is, instead of using a %, you insert a pair of curly brackets where you want your variable values to be. Then, you follow the string with the `.format()` method. The `.format()` method has several advantages over the %s placeholder method. These are illustrated in the following code:

```
1  '''
2  Examples of the .format method
3  '''
4  # 1. Inserted objects can be called by index position:
5  print('The {2} {1} {0}'.format('fox','brown','quick'))
6
7  # 2. Inserted objects can be assigned keywords:
8  print('First Object: {a}, Second Object: {b}, Third Object: \
9       {c}'.format(a=1,b='Two',c=12.3))
10
11 # 3. Inserted objects can be reused, avoiding duplication:
12 print('A {p} saved is a {p} earned.'.format(p='penny'))
13
14 # You can do aligning, padding and precision with .format()!
15 print('{0:8} | {1:9}'.format('Fruit', 'Quantity'))
16 print('{0:8} | {1:9}'.format('Apples', 3.))
17 print('{0:8} | {1:9}'.format('Oranges', 10))
18
19 # By default, .format() aligns text to the left,
20 # numbers to the right.
21 # You can pass an optional <,^, or > to set a left, center or
22 # right alignment:
23 print('{0:<8} | {1:^8} | {2:>8}'.format('Left','Center','Right'))
24 print('{0:<8} | {1:^8} | {2:>8}'.format(11,22,33))
25
26 # You can precede the alignment operator with a padding character
27 print('{0:=<8} | {1:-^8} | {2:.>8}'.format('Left','Center',
28      'Right'))
29 print('{0:=<8} | {1:-^8} | {2:.>8}'.format(11,22,33))
30
31 # Field widths and float precision are handled in a way similar
32 # to placeholders.
33 # The following two print statements are equivalent:
34 print('This is my ten-character, two-decimal number:
35 %10.2f' %13.579)
36 print('This is my ten-character, two-decimal \
37      number:{0:10.2f}'.format(13.579))
```

Listing 1.8: Sample code showing how to use the .format method.

This results in the output:
```
The quick brown fox
First Object: 1, Second Object: Two, Third Object: 12.3
A penny saved is a penny earned.
Fruit      | Quantity
Apples     |      3.0
Oranges    |       10
Left       | Center  |   Right
11         |   22    |      33
Left====   | Center  | ...Right
11======   | ---22--- | ......33
This is my ten-character, two-decimal number:     13.58
This is my ten-character, two-decimal number:     13.58
```

So much for formatted output to the screen; now on to input from the keyboard. Fortunately, this is very simple to do. Here is a code snippet that demonstrates input from the keyboard:

```
1  your_cat = input("What is your cat's name? ")
2  print("Your cat's name is {}?  How cute!".format(your_cat))
3
4  # Multiple inputs:
5  dogs_names= input('I hear you have two dogs. \
6              Input their names, separated by a comma: ")
7  list_of_names = [s.strip() for s in dogs_names.split(",")]
8  print("Your dogs names are {} and {}. \
9  Not very original!".format(list_of_names[0],list_of_names[1]))
10 # Inputting non-strings:
11 favorite_number = float(input("What's your favorite number? ")) \\
12 print("You said your favorite number is {}".format(favorite_
13     number))
```
Listing 1.9: Sample code showing how to use the **input** command.

If you execute this code, You will see the output below, where the items in square brackets are presumably what you typed in response to the questions. The code snippet then uses your input, in this case by simply printing it out. In line 1 we are looking for a single input. Lines 5-9, we show how to deal with multiple inputs. (Note, by the way, that because lines 5 and 8 were too long, I broke it up using the \character.) This works because the variable on the left-hand side of the **input** statement is assumed to be of type string. Thus, line 6 simply parses this string into elements of a list, which are printed out in lines 7-8.

```
What is your cat's name? [Whiskers]
Your cat's name is Whiskers? How cute!
I hear you have two dogs. Input their names, separated
by a comma: [Fido,Spot]
Your dogs' names are Fido and Spot. Not very original!
What's your favorite number? [3.14159]
You said your favorite number is 3.14159
```

TABLE 1.1: File read/write indicators

Mode	Descsription
r	opens a file for reading only
w	opens a file for writing. If the file exists, it overwrites it. Otherwise, it creates a new file.
a	opens a file for appending only. If the file doesn't exist, it creates the file.
x	creates a new file. If the file exists, it fails.
+	opens a file for updating.

This is all simple enough, but what if you'd like your input to be, for example, of type **float**? Simple! Just use the **float** command to convert the output of **input** to a variable of type float. This is shown in line 11.

1.4.2 Input/Output to/from Files

Now we're ready to see how to write text and numbers to a file. This is very important because when you are doing some sort of numerical computation, you generally want to keep your data, rather than have it fly off the screen. Here's a sample code snippet wherein we write to a file:

```python
'''
Outputting data to a file
'''
# use a 'w' to tell Python you want to WRITE to the specified
    file:
f = open('output.txt','w')

# note the use of '\n', the new line symbol:
f.write('This will be the first line of the file.\n')
f.write("{} {} at ${:4.2f} each\n".format(20, 'liters', 1.17234))
f.close()
```

Listing 1.10: Sample code showing how to open and write to a file.

Line 5 opens a file named "output.txt". As the comment indicates, the file will be available to write to. In this statement, we also define a variable, **f**, which now represents that file. So, in all future interactions with that file, we refer to it via that variable. You can see this in lines 8-10.

Besides using "**w**" to write to a file, we have other options; these apply to both reading and writing files. A summary is in Table 1.1.

Now let's see how we write to this file we've created. As you can see in lines 8 and 9, we simply substitute "**f.write**" for "**print**, where "**f**" is whatever variable name we have chosen to represent that file. I've used the **format** method here, but any formatting option you choose will work. Note that I wrote both text and numbers to the file. This is typical because it is usually helpful to generate column headings for multi-column data. Line 10 is very important. You must make sure you close the file once you've finished writing to it in order to be sure that the last bit of data has been sent to it.

If I were to look in the file I just created, I would see:

This will be the first line of the file.
20 liters at \$1.17 each

This is pretty much all there is to writing to a file. The only hard part is mastering formatting, and as we've already said, the formatting works the same way as when writing to the screen.

Now we need to learn how to read from a file. There are several ways to do this. Let's start with a text file. Suppose I have a file of text called "This_Old_Man.txt", which contains the lyrics of the first stanza of that song. I can read the file and print it out, line by line, with the following code.

```
1  '''
2  Reading a text file line by line
3  '''
4  with open('This_Old_Man.txt') as f:
5      line = f.readline()
6      while line:
7          print(line, end='')
8          line = f.readline()
9  print()     # just to create a space between the two outputs...
10
11 # another way to read a text file:
12 with open('This_Old_Man.txt') as f:
13     lines = f.readlines()
14 print(lines)
```

Listing 1.11: Sample code showing how to open and read a text file.

If I execute this code I get on my screen,

This old man, he played one,
He played knick-knack on my thumb;
With a knick-knack paddywhack,
Give the dog a bone,
This old man came rolling home.

['This old man, he played one,\n', 'He played knick-knack on my
thumb;\n',
'With a knick-knack paddywhack,\n', 'Give the dog a bone,\n',
'This old man came rolling home.\n']

Let's examine the code to see what's going on. In line 4 we open the file, assigning to it the *object* (what we've been calling a file variable) **f**. Using the keyword **with** we set up a loop. (Note the colon and the indentation.) The first line in the block, line 5, uses the **readline** method. This reads a single line of code and stores it in the variable **line**. If the read is successful, this assignment will be **True**. For the rest of the loop, lines 6-8, have a sub-loop, a **while**-loop. This prints out the current value of **line**, and reads another line. The **while**-loop continues to read and print as long as the assignment of the variable **line** is **True**. It will fail (return a **False** when the readline method returns an empty string. This will end the **while**-loop.

But what if the file we're trying to read is not one big string? What if it contains columns of numbers? Well, here is some code that shows us one way to deal with this. First, let's look at the file I want to read. The name of the file is "data1.dat" and it contains:

```
0    1
1    3
2    5
3    7
4    9
5    11
```

```
1  '''
2  Reading  columns  of  data  from  a  file
3  '''
4  with open('data1.dat') as f:
5      lines = f.readlines()
6
7  x = []
8  y = []
9  for i in lines:
10      piece = i.split("   ")
11      x.append(float(piece[0]))
12      y.append(float(piece[1]))
13  print(x)
14  print(y)
```

Listing 1.12: Sample code showing how to open and read a file containing columns of data.

When I execute this code, I get
[0.0, 1.0, 2.0, 3.0, 4.0, 5.0]
[1.0, 3.0, 5.0, 7.0, 9.0, 11.0]

Notice that the left column is placed in a list named x and the right column is placed in a list named y. This format makes the data easy to process. Let's look at the code and see how it works. We open and read the file as before, as shown in lines 4 and 5. Upon completion of line 5, the variable **lines** will be a list of strings, with each element of **lines** containing one line from the data file. Then we create two empty lists, one for x and one for y. Finally, we use the **split** method to extract the x and y values from each element of **lines**. Note the use of **float** to convert each piece of the string into a floating point number. Also note that the argument of the **split** method contains whatever character(s) were used to delimit the numbers in the file, in this case, 3 spaces. In Python it is very often the case that one starts with strings or lists of strings, and strips them to get to the data.

There are other ways to read data files, csv files, *etc*. But this should get us started. I recommend that you start with just a few simple pieces of code that you can use as templates. Once you're comfortable with them, you can start thinking about using other approaches for file I/O.

1.5   FUNCTIONS

Python comes with many built-in functions. We've already used some of them, and many more are contained in the numerical packages we'll discuss later in this chapter. But it very often the case that you'll want to create your own functions. In Python this is very easy to do. Here is a trivial example.

```
1  '''
2  Defining  Functions
3  '''
4  def  myfct(s):
5      temp = 3.0*s + 7.5
6      return  temp
7
8  x = []
9  y = []
10 for  i  in  range(10):
11     x.append(float(i))
12 print(x)
13
14 for  i  in  x:
15     y.append(myfct(i))
16 print(y)
```

Listing 1.13: Sample code showing how to create and use a function.

When I execute this code, I get
$[0.0, 1.0, 2.0, 3.0, 4.0, 5.0, 6.0, 7.0, 8.0, 9.0]$
$[7.5, 10.5, 13.5, 16.5, 19.5, 22.5, 25.5, 28.5, 31.5, 34.5]$

As you can see, creating and using functions is very easy. One caveat is that the function must be defined *before* its first use. Notice that the argument in the function definition need not match the argument when you call the function. In fact all the variables in the function definition are *local* to that definition block. A function can be as complicated or as simple as you like, and it can have multiple arguments.

1.5.1   MatPlotLib

Many powerful – and free – packages are available for Python. These greatly expand the capabilities of this already powerful language. In this section we will discuss one of them, `matplotlib`, a powerful and convenient plotting package. It comes bundled with the Anaconda implementation, but can be separately installed if you've installed Python in a different way.

The first thing we need to do is to **load** the package. Then we plot!

```
1  '''
2  Simple line plot using MatPlotLib
3  '''
4  from matplotlib import pyplot as plt
5  import numpy as np
6  %matplotlib inline
7
8  plt.figure()
9  plt.plot([1,2,3,4])
10 plt.show()
```

Listing 1.14: Sample code showing how to create a basic line plot

First look at line 4. Here we import the MatPlotLib package. You see, even though Anaconda comes bundled with many add-on packages, it will only load them for your application if you say so. This way the program does not occupy as much space in your computer's memory. Still in line 4, you can see that we only imported the **pyplot** sub-package within MatPlotLib. Again, this reduces the space your program would occupy on your computer compared to if you had loaded in the entirety of the MatPlotLib package. Finally, in line 4 we tell Python that whenever we refer to a function or method in MatPlotLib, we will preface that method with "**plt**". Note that we could have called this anything, "**gorp**" for example. But **plt** is more descriptive and, as it turns out, is the name typically used by most Python users. But why use any descriptor at all? In general, when you load in a package, you'll want to use a descriptor because it is possible that in your application, you may create your own functions. Now what if you inadvertently give your function that same name that exists in the package you've loaded? This is called a namespace conflict and can lead to unexpected – and difficult to diagnose – consequences. Ok, on to line 5 where we load the numpy package, which stands for numerical Python. Technically, we do not need numpy here, but for practically any numerical computation you will. So my recommendation is to load it for every program you write. Now, line 6. This is a so-called "*magic*" command and is only needed if you are doing your plotting from within a Jupyter Notebook. If you are implementing your code outside of Jupyter, you do not need this command. With lines 8-10 we plot. Line 8 basically alerts MatPlotLib that we are going to make a plot. There can be an argument in the **figure()** method. This can be handy if you are making multiple plots and need to tell MatPlotLib which one you are referring to. We'll see this later. Next we have the command to create the plot. At the same time, we created the data we want to plot. This is not the normal way to do things! Normally, you would create your x- and y-data, put them into a list, and then tell MatPlotLib to plot the data. By default, the plot will be a line plot, but as we'll see, you can also make this a point plot, with or without error bars, or any combination of these options. The default color is blue, but you can control the colors as well. Notice that we did not create axis labels or a plot title – but we could have. Now, nothing will be plotted until Python sees the **show()** method. The result is shown in Fig. 1.1.

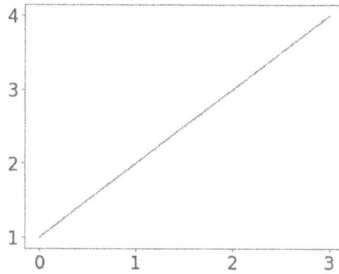

FIGURE 1.1: Our first MatPlotLib plot!

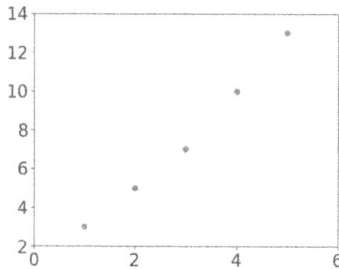

FIGURE 1.2: Simple point plot with axis labels and a title.

Here is a more usual way to plot data, along with the output:

```
1  '''
2  Simple point plot using MatPlotLib
3  '''
4  from matplotlib import pyplot as plt
5  import numpy as np
6  %matplotlib inline
7
8  x = [1,2,3,4,5]
9  y = [3,5,7,10,13]
10 plt.figure(1)
11 plt.axis([0,6,2,14])
12 plt.xlabel('time (sec)')
13 plt.ylabel('position (cm)')
14 plt.title('Simple Point Plot')
15 plt.plot(x,y,'ro')
16 plt.show()
```

Listing 1.15: Sample code showing how to make a simple point plot

In Listing 1.15 we first loaded in the MatPlotLib package, then executed the magic command. In lines 8 and 9 we created our data. Lines 10 and 16 are as before, but lines 11-15 are new. In line 11 instead of letting MatPlotLib decide on the limits

TABLE 1.2: Defining line types and colors

Character	Descsription
'-'	solid line
'- '	dashed line
'-.'	dashed-dot line
':'	dotted line
'.'	point marker
','	pixel marker
'o'	circle marker
's'	square marker
'^'	triangle marker
'v'	inverted triangle marker
'*'	star marker
'+'	plus marker
'D'	diamond marker
'd'	thin diamond marker

Character	Color
'b'	blue
'g'	green
'r'	red
'k'	black
'w'	white
'c'	cyan
'm'	magenta
'y'	yellow

to the x- and y-axis, we did so, manually. MatPlotLib usually does a good job, but sometimes you will find the need to tweak a bit. In lines 12-14 we created axis labels and a graph title. Those commands are straight-forward. But now look at line 16. We still used the `plot` method, but we've used it somewhat differently from before. First, we inserted the names of our lists, rather than the data itself. (Always give the horizontal axis data first.) We've also added an extra argument, a string that tells MatPlotLib that we want the graph to be red ('r') and that we want a point plot, with the points represented by solid circles. You actually have great latitude in selecting colors and point (or line) types. Table 1.2 gives a partial summary of these.

Getting into all the capabilities of MatPlotLib is beyond the scope of this chapter on Python. But there are fabulous books and on-line resources on this topic.

1.6 NUMERICAL PACKAGES: NUMPY AND SCIPY

Both of these packages have many, many built-in functions and classes, and we will only cover a small number of them, with the idea of giving you the tools to numerically address the math topics covered in the rest of this text. But even the limited discussion of these packages will provide you with numerically capabilities that go far beyond the math topics covered here.

Lets start with **numpy**, which stands for "numerical python". This package, which typically comes bundled with the main Python package, contains commonly used functions and physical constants, as well as the **ndarray** class, tools for linear algebra (Chapter 3) and tools for Fourier transforms (Chapter 5). **ndarray** is a much more efficient way to manipulate arrays than Python lists. You'll see **ndarray** in action in Chapter 3.

You will see many other examples of **numpy** throughout the rest of this text. Here is a simple example, showing a common way to load **numpy** and how to use it to obtain the value of π.

```
1  '''
2  Loading and using numpy
3  '''
4  import numpy as np
5
6  print(np.pi)
```

Listing 1.16: Sample code showing how to load **numpy** and use it to get the value of π

Take special note of how we loaded in the **numpy** package. We followed exactly this protocol when we loaded in **matplotlib**, but used the conventional "np" to reference calls to **numpy**. We then used the **pi** method to obtain the stored value of π from the **numpy** package.

The **scipy** package is actually a superset of **numpy**. That is, all of **numpy** is contained in **scipy**, so if you load the **scipy** package, you need not load **numpy**. **scipy** stands for "scientific python". In addition to the contents of **numpy**, **scipy** also contains functions for optimization, statistical packages, and functions for signal processing. For almost any scientific applications using Python, you will load **numpy**; but you may or may not need to load **scipy**.

## CHAPTER 2

Introduction to Complex Algebra

2.1 THE BASICS

In nearly all areas of physics we will find it essential to be able to manipulate complex quantities. For example, quantum mechanical wave functions are generally complex quantities. In electronics, the impedances of some "reactive" elements, such as inductors and capacitors, are best described using complex algebra. It is safe to say you will not understand physics, beyond Newton's Laws, without being comfortable with complex algebra.

This chapter contains a brief overview of manipulating complex numbers. First off, what *is* a complex number (or variable)? Simply stated, a complex number is a number that contains a "pure real" part and a "pure imaginary" part. We'll explain what this means as we work through this chapter.

Let's suppose z is a complex number. Then we can express z as the sum of a pure real and a pure imaginary number:

$$z = a + ib, \tag{2.1}$$

where a and b are pure real and $i \equiv \sqrt{-1}$. (Note that the electronics world uses j instead of i for the square root of -1. The reason is that in electronics the symbol i is generally used to designate a current. In this text, we will stick with i, not j, as the symbol representing $\sqrt{-1}$).

The complex conjugate of a complex number is defined as replacing i by $-i$. Notationally, the complex conjugate of z is expressed as z^*. Thus the complex conjugate of z from Eq. 2.1 is

$$z^* = a - ib. \tag{2.2}$$

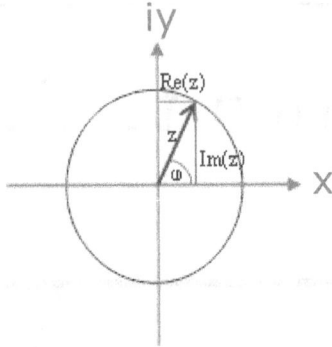

FIGURE 2.1: Graphical representation of the complex number z.

Then, using Eqs. 2.1 and 2.2,

$$Re(z) = \frac{(z + z^*)}{2},$$ (2.3a)

$$Im(z) = \frac{(z - z^*)}{2i},$$ (2.3b)

where $Re(z)$ and $Im(z)$ are the real and imaginary parts (a and b)of z, respectively.

Complex numbers can also be expressed in terms of a magnitude and a phase. The magnitude of complex number z is given by

$$|z| = \sqrt{z^* z}$$ (2.4a)

$$= \sqrt{(a - ib)(a + ib)}$$ (2.4b)

$$= \sqrt{a^2 + b^2}$$ (2.4c)

It is often useful to think of complex numbers graphically as lying in the complex $x - iy$ plane. For example, consider the number z in the complex plane, as shown in Fig. 2.1.

If, as in Eq. 2.1, the real and imaginary parts of z are a and b, respectively, then from Pythagoras' theorem, we obtain Eq. 2.4c. Furthermore, the angle ϕ is given by

$$\phi = \tan^{-1}(b/a),$$ (2.5)

and is referred to as the phase. We can then express z as

$$z = |z|e^{i\phi}.$$ (2.6)

Recalling Euler's theorem,

$$e^{i\phi} = \cos(\phi) + i\sin(\phi) ,$$ (2.7)

and looking at Fig. 2.1, we can see that $a = |z|\cos\phi$ and $b = |z|\sin\phi$.

Ok, we can see what ϕ "means" in Fig. 2.1, but what does this phase mean *physically*? Here's an example: Suppose you have an electromagnetic wave $E_0\cos(\omega t)$.

Notice I did not include a phase – yet. Now suppose this electromagetic field passes through a material having an index of refraction, n. As you will learn in a course on electromagnetic theory, the wave will travel more slowly through this material than through vacuum. Thus, the wave passing through the material will be delayed, compared to an identical wave that does not pass through that material. That delay, when expressed as an angle, is the phase shift of the wave that had passed through the material, compared to the wave that did not. So how to convert that time shift into an angle? Well, one full cycle corresponds to an angle of 2π radians, or 360 degrees. For example, if the temporal delay between two waves, both having period T, is Δt, then the phase difference, ϕ between the two waves is given by:

$$\frac{\Delta t}{T} = \frac{\phi}{2\pi}, \text{ or}$$

$$\phi = 2\pi \frac{\Delta t}{T}. \tag{2.8}$$

Generally speaking, Eq. 2.1 is often the more useful representation for manipulating complex numbers, while Eq. 2.6 is often more useful in their interpretation. For example, in quantum mechanics, wave functions are usually described in terms of a magnitude and phase.

Homework 2.1: Manipulating Complex Quantities

In an RC high-pass filter, the ratio of the output voltage V_{out} to the input voltage V_{in}, is given by

$$\frac{Vout}{Vin} = \frac{R}{R + Z_C} \tag{2.9}$$

where R is the resistivity of the resistor and Z_C is the complex impedance of a capacitor, given by

$$Z_C = \frac{1}{i\omega C}. \tag{2.10}$$

Here, $\omega = 2\pi f$, where f is the frequency of the signal seen by the capacitor, and C is the capacitance of the capacitor. *Don't worry if you do not understand the physics in this problem! You will learn this when you study electronics.*

Inserting Eq. 2.10 into Eq. 2.9, show how to express the output to input ratio in the form

$$\frac{V_{out}}{V_{in}} = |S|e^{i\phi}, \tag{2.11}$$

where

$$|S| = \frac{\omega RC}{\sqrt{1 + \omega^2 R^2 C^2}} \quad \text{and}$$

$$\phi = \tan^{-1}\left(\frac{1}{\omega RC}\right). \tag{2.12}$$

Swell, but what does this solution actually mean? Well, from the expression for $|S|$, we can see that the magnitude of the output signal is smaller than that of the input signal by a quantity that depends on the values of the resistor, the capacitor, and the frequency of the signal. For example, for the specific case in which

$$\omega = \frac{1}{RC} \tag{2.13}$$

we can see that the magnitude of the output will be reduced by a factor of $1/\sqrt{2}$, compared to the input. This is the famous -3 dB point. What about the phase, ϕ? If the input is a sine wave having angular frequency ω, then the output will also be a sine wave at that frequency. But, the output sine wave will be shifted by an "angle" given by 2.12. For example, if we have the condition of Eq. 2.13, the phase would be

$$\phi = \tan^{-1}(1) = \frac{\pi}{4}\text{radians} = 45°$$

In the case of a frequency at the 3dB point (Eq. 2.13), the output would be temporally shifted by 1/8 of the period. The point of this exercise is twofold: practice in manipulating complex variables *and* seeing a real world interpretation of a complex expression.

2.2 USING PYTHON WITH COMPLEX NUMBERS

Python would lose much of its utility in scientific programming if it could not conveniently work with complex numbers. In this short section, we show a few of the built-in methods that you can use to manipulate complex quantities.

```python
'''
Complex Methods
'''
import numpy as np
z1 = 1 + 2j
z2 = -3 + 1j
print("Re(z1) = ", z1.real)
print("Im(z1) = ", z1.imag)
print("phase(z1) = ", np.angle(z1))
print("z1 x z2 = ", z1*z2)
print("|z1 x z2| = ", np.absolute(z1*z2))
print()
# Also works on arrays!
a = np.array([1+2j, 3+4j, 5+6j])
b = np.conj(a)
angle_a = np.angle(a)

print("a= ", a)
print("Re(a) = ", a.real)
print("Imag(a) = ", a.imag)
print("complex conjugate(a) == a* =", b)
print("|a| = ", abs(a))    # an alternate way of obtaining the \
        magnitude
print("phase(a) = ", angle_a)
```

Listing 2.1: Sample code showing how to compute with complex numbers.

We'll talk more about arrays (lines 14-24) in the next chapter. For now, think of them as extra-powerful lists. Note, by the way, that in Python, j (rather than i) is used to represent $\sqrt{-1}$. When I execute this code, I get

Re(z1) = 1.0
Im(z1) = 2.0
phase(z1) = 1.1071487177940904
z1 x z2 = (-5-5j)

|z1 x z2| = 7.0710678118654755
a= [1.+2.j 3.+4.j 5.+6.j]
Re(a) = [1. 3. 5.]
Imag(a) = [2. 4. 6.]
complex conjugate(a) == a* = [1.-2.j 3.-4.j 5.-6.j]
|a| = [2.23606798 5. 7.81024968]
phase(a) = [1.10714872 0.92729522 0.87605805]

All of these results are what we should expect from what we've learned in the beginning part of this chapter. So, the point of this section is, that if you are deciding whether or not to use Python or "manual" manipulation of variables, the fact that your variables may be complex should not even enter the discussion; Python works equally well – and conveniently – for integers, floating reals, and complex numbers.

## CHAPTER 3

Matrices

3.1 BASIC MATRIX OPERATIONS

3.1.1 What is a Matrix?

A matrix is an array of numbers, symbols, or expressions, arranged in rows and columns. Working with these can be extremely powerful for solving problems in math, physics, and engineering.

Consider the $n \times m$ matrix A, where n is the number of rows and m is the number of columns. We can designate any matrix element by a_{ij}, where i is the row number and j is the column number of that element. For example, a generic $n \times m$ matrix A looks like

$$A = \begin{pmatrix} a_{11} & a_{12} & a_{13} & \cdots & a_{1m} \\ a_{21} & a_{22} & & & \\ a_{31} & & \ddots & & \\ \vdots & & & & \\ a_{n1} & & & & \end{pmatrix}. \tag{3.1}$$

There are three basic matrix operations: matrix addition, multiplication of a matrix by a scalar, and the multiplication of a matrix by another matrix. We will also talk about the *transpose* of a matrix, designated by A^T, and the inverse of a matrix, designated by A^{-1}.

To transpose a matrix you swap row for column: if $B = A^T$, then $b_{ij} = a_{ji}$ for all i, j. For example, for

$$A = \begin{pmatrix} 1 & 2 \\ 3 & 4 \\ 5 & 6 \end{pmatrix} \tag{3.2a}$$

$$B = A^T = \begin{pmatrix} 1 & 3 & 5 \\ 2 & 4 & 6 \end{pmatrix} \tag{3.2b}$$

3.1.2 Multiplication by a Scalar

When we multiply a matrix by a scalar, we multiply every element in the matrix by that number. Thus, $tA = ta_{ij}$ for every a_{ij} in matrix A.

3.1.2.1 Example of Multiplication by a Scalar

$$\text{if } A = \begin{pmatrix} 1 & 4 \\ 2 & 2 \end{pmatrix} \quad \text{then} \tag{3.3a}$$

$$5A = \begin{pmatrix} 5 & 20 \\ 10 & 10 \end{pmatrix} \tag{3.3b}$$

There is no restriction on the dimensionality of a matrix for multiplication by a scalar to be valid. Furthermore, matrix multiplication by a scalar is *commutative*. That is, $tA = At$.

3.1.3 Matrix Addition

Consider two matrices A and B:

$$A = \begin{pmatrix} a_{11} & a_{12} \\ a_{21} & a_{22} \end{pmatrix} \tag{3.4a}$$

$$B = \begin{pmatrix} b_{11} & b_{12} \\ b_{21} & b_{22} \end{pmatrix} \tag{3.4b}$$

We can *add* these matrices by adding the elements of one matrix to the corresponding elements of the other matrix. Thus,

$$A + B = \begin{pmatrix} a_{11} + b_{11} & a_{12} + b_{12} \\ a_{21} + b_{21} & a_{22} + b_{22} \end{pmatrix} \tag{3.5}$$

Note that matrix addition is also commutative, *i.e.* $A+B = B+A$. Also note that for matrix addition to be valid, the two matrices must have the same dimensionality $(n \times m)$. As an example, if

$$A = \begin{pmatrix} 1 & 3 & -5 \\ 0 & 2 & 6 \end{pmatrix} \quad \text{and} \tag{3.6a}$$

$$B = \begin{pmatrix} -3 & 2 & 4 \\ 6 & -1 & 7 \end{pmatrix} \quad \text{then} \tag{3.6b}$$

$$A + B = \begin{pmatrix} -2 & 5 & -1 \\ 6 & 1 & 13 \end{pmatrix} \tag{3.6c}$$

You can think of matrix subtraction as first multiplying one matrix by -1, and then adding that to the other matrix.

$$A - B = \begin{pmatrix} a_{11} - b_{11} & a_{12} - b_{12} \\ a_{21} - b_{21} & a_{22} - b_{22} \end{pmatrix} \tag{3.7}$$

3.1.4 Matrix Multiplication

Now things get a bit complicated. To multiply one matrix by another, you take the entire top *row* of the left matrix and multiply it, element by element, by the first *column* of the other matrix, and add all those products. You then repeat this for all rows and columns. Thus, for the ij elements of the product $n \times m$ matrix,

$$(ab)_{ij} = \sum_{k=1}^{m} a_{ik} b_{kj} \qquad (3.8)$$

Note that for matrix multiplication to be possible, the number of columns in the left-hand matrix must equal the number of rows in the right-hand matrix. In general, matrix multiplication does not commute.

3.1.4.1 Example of Matrix Multiplication

$$A = \begin{pmatrix} 1 & 3 & -5 \\ 0 & 2 & 6 \end{pmatrix} \qquad (3.9a)$$

$$B = \begin{pmatrix} -3 & 6 \\ 2 & -1 \\ 4 & 7 \end{pmatrix} \qquad (3.9b)$$

$$AB = \begin{pmatrix} (-3+6-20) & (6-3-35) \\ (0+4+24) & (0-2+42) \end{pmatrix} \qquad (3.9c)$$

$$= \begin{pmatrix} -17 & -32 \\ 28 & 40 \end{pmatrix} \qquad (3.9d)$$

Homework 3.1: Multiplying Matrices

For

$$A = \begin{pmatrix} 1 & 2 & 3 \\ 4 & 5 & 6 \end{pmatrix} \; ; \; B = \begin{pmatrix} 2 & 4 \\ 6 & 8 \\ 10 & 12 \end{pmatrix} \qquad (3.10)$$

find AB and BA.

3.1.5 Matrix Division

Suppose you had the equation $ab = c$, where a, b, and c all just represent numbers, and you wanted to find what b was. You could simply divide both sides by a and obtain $b = c/a$. Another way to think about it is that you can multiply both sides of the equation by the inverse of a, a^{-1}, to get the answer.

When dealing with matrices, we don't have a way to directly divide, so we have to multiply by the inverse. However, finding this inverse can sometimes be tricky. **You cannot just invert all the elements**. In other words, if

$$A = \begin{pmatrix} 5 & 5 \\ 3 & 1 \end{pmatrix} \qquad (3.11)$$

then

$$A^{-1} \neq \begin{pmatrix} 1/5 & 1/5 \\ 1/3 & 1 \end{pmatrix} \tag{3.12}$$

3.1.6 Matrix Inversion

By definition,

$$A^{-1}A = AA^{-1} = I_N \tag{3.13}$$

where I is the *unity matrix*, and the (optional) subscript refers to its dimension. The unity matrix has zeroes for all of its elements except the diagonal, which is filled with 1's. For example, the 3×3 unity matrix is:

$$I_3 = \begin{pmatrix} 1 & 0 & 0 \\ 0 & 1 & 0 \\ 0 & 0 & 1 \end{pmatrix} \tag{3.14}$$

3.1.6.1 Example of an Inverse Matrix

For

$$A = \begin{pmatrix} 5 & 2 \\ 3 & 1 \end{pmatrix} \tag{3.15a}$$

$$A^{-1} = \begin{pmatrix} -1 & 2 \\ 3 & -5 \end{pmatrix} \tag{3.15b}$$

we can verify that the matrix in Eq. 3.15b is the inverse of the matrix in Eq. 3.15a multiplying AA^{-1}:

$$AA^{-1} = \begin{pmatrix} 5 & 2 \\ 3 & 1 \end{pmatrix} \begin{pmatrix} -1 & 2 \\ 3 & -5 \end{pmatrix} = \begin{pmatrix} 1 & 0 \\ 0 & 1 \end{pmatrix} \tag{3.16}$$

and

$$A^{-1}A = \begin{pmatrix} -1 & 2 \\ 3 & -5 \end{pmatrix} \begin{pmatrix} 5 & 2 \\ 3 & 1 \end{pmatrix} = \begin{pmatrix} 1 & 0 \\ 0 & 1 \end{pmatrix} \tag{3.17}$$

3.1.6.2 How to Compute the Inverse of a Matrix

If the inverse of a matrix exists, it can be computed. First we build an *augmented matrix* of the form shown below. The left part is the matrix which is to be inverted, and the right part is the unity matrix of matching dimension.

$$\begin{pmatrix} 5 & 2 & | & 1 & 0 \\ 3 & 1 & | & 0 & 1 \end{pmatrix} \tag{3.18}$$

The inverse of A above can be found by *row-reducing* the left-hand side until it looks like the unity matrix. This is called Gauss elimination. In order to do this you have to perform one of the following operations to the augmented matrix:

- You can multiply every element (including both sides of the vertical bar) in a row by a non-zero scalar.
- You can exchange any two rows.
- You can add a scalar multiple of one row to another row.

Here's an example:

$$\left(\begin{array}{cc|cc} 5 & 2 & 1 & 0 \\ 3 & 1 & 0 & 1 \end{array} \right) \Rightarrow_{R_1/5 \to R_1} \left(\begin{array}{cc|cc} 1 & \frac{2}{5} & \frac{1}{5} & 0 \\ 3 & 1 & 0 & 1 \end{array} \right) \tag{3.19}$$

Here $R_1/5 \to R_1$ means "divide row 1 by 5 and put the result in place of the original row 1". We are trying to get a 1 in the top left corner.

$$\left(\begin{array}{cc|cc} 1 & \frac{2}{5} & \frac{1}{5} & 0 \\ 3 & 1 & 0 & 1 \end{array} \right) \Rightarrow_{3R_1 - R_2 \to R_2} \left(\begin{array}{cc|cc} 1 & \frac{2}{5} & \frac{1}{5} & 0 \\ 0 & \frac{1}{5} & \frac{3}{5} & -1 \end{array} \right) \tag{3.20}$$

Here $3R_1 - R_2 \to R_2$ means "multiply row 1 by 3 and then subtract row 2 from that. Put the result in place of the original row 2." We do this to get a 0 in the bottom left corner.

$$\left(\begin{array}{cc|cc} 1 & \frac{2}{5} & \frac{1}{5} & 0 \\ 0 & \frac{1}{5} & \frac{3}{5} & -1 \end{array} \right) \Rightarrow_{R_1 - 2R_2 \to R_1} \left(\begin{array}{cc|cc} 1 & 0 & -1 & 2 \\ 0 & \frac{1}{5} & \frac{3}{5} & -1 \end{array} \right) \tag{3.21}$$

Now there is a 0 in the upper right corner of the left-hand side.

$$\left(\begin{array}{cc|cc} 1 & 0 & -1 & 2 \\ 0 & \frac{1}{5} & \frac{3}{5} & -1 \end{array} \right) \Rightarrow_{5R_2 \to R_2} \left(\begin{array}{cc|cc} 1 & 0 & -1 & 2 \\ 0 & 1 & 3 & -5 \end{array} \right) \tag{3.22}$$

The matrix on the right side is the inverse of the matrix we had originally placed on the left side.

Homework 3.2: Computing the inverse of a matrix

If

$$A = \left(\begin{array}{cc} 3 & 1 \\ 4 & 2 \end{array} \right) \tag{3.23}$$

find A^{-1}. Verify that it is, indeed, the inverse by showing the products AA^{-1} and $A^{-1}A$ equal the identity matrix.

3.2 USING MATRICES TO SOLVE SYSTEMS OF LINEAR EQUA-TIONS

Consider the system of equations:

$$2x + 3y + z = 11 \tag{3.24a}$$
$$y - 2z = -4 \tag{3.24b}$$
$$x - 4y - z = -10 \tag{3.24c}$$

The first step is to write this as a matrix multiplication equation, $A\vec{x} = \vec{d}$, where A is a matrix which contains the coefficients in front of the x, y, and z. $\vec{x}$ is a column matrix of x, y, and z. You can think of this as a position vector, though it can be written in as many dimensions as desired. Finally, $\vec{d}$ is a column matrix containing the numbers on the right-hand side of the equations.

Keep in mind that this works for the x, y, z of the Cartesian coordinate system, but can be generalized to any quantity of variables. As a result, it is common to generalize using the variables x_i, such that the $\vec{x}$ matrix would be a column of x_1, x_2, ...

For the example of Eqs. 3.24, $A\vec{x} = \vec{d}$ would look like,

$$\begin{pmatrix} 2 & 3 & 1 \\ 0 & 1 & -2 \\ 1 & -4 & -1 \end{pmatrix} \begin{pmatrix} x \\ y \\ z \end{pmatrix} = \begin{pmatrix} 11 \\ -4 \\ -10 \end{pmatrix} \tag{3.25}$$

You can see that if you actually perform this multiplication, you will get the original Eqs. 3.24 back.

We will then represent this matrix equation by an augmented matrix. To solve the system of equations represented by the augumented matrix, we will row-reduce the augmented matrix to the unity matrix as we did when we inverted a matrix.

When solving, because we don't want to write out the $\vec{x}$ matrix every time explicitly, we use a more simplified notation which replaces the "$\vec{x}$" with the vertical line we usually see in augmented matrices:

$$\left(\begin{array}{ccc|c} 2 & 3 & 1 & 11 \\ 0 & 1 & -2 & -4 \\ 1 & -4 & -1 & -10 \end{array} \right) \tag{3.26}$$

We now proceed to row-reduce as we did in the matrix inversion process:

$$\left(\begin{array}{ccc|c} 2 & 3 & 1 & 11 \\ 0 & 1 & -2 & -4 \\ 1 & -4 & -1 & -10 \end{array} \right) \Rightarrow_{R_1 \leftrightarrow R_3} \left(\begin{array}{ccc|c} 1 & -4 & -1 & -10 \\ 0 & 1 & -2 & -4 \\ 2 & 3 & 1 & 11 \end{array} \right) \tag{3.27}$$

where $R_1 \leftrightarrow R_3$ means to swap rows 1 and 3. Continuing,

$$\left(\begin{array}{ccc|c} 1 & -4 & -1 & -10 \\ 0 & 1 & -2 & -4 \\ 2 & 3 & 1 & 11 \end{array} \right) \Rightarrow_{2R_1 - R_3 \to R_3} \left(\begin{array}{ccc|c} 1 & -4 & -1 & -10 \\ 0 & 1 & -2 & -4 \\ 0 & -11 & -3 & -31 \end{array} \right) \tag{3.28}$$

Here, $2R_1 - R_3 \to R_3$ means "take 2 times row 1, subtract row 3 from that, and put the result in place of the original row 3".

$$\left(\begin{array}{ccc|c} 1 & -4 & -1 & -10 \\ 0 & 1 & -2 & -4 \\ 0 & -11 & -3 & -31 \end{array} \right) \Rightarrow_{11R_2 + R_3 \to R_3} \left(\begin{array}{ccc|c} 1 & -4 & -1 & -10 \\ 0 & 1 & -2 & -4 \\ 0 & 0 & -25 & -75 \end{array} \right) \tag{3.29}$$

$$\left(\begin{array}{ccc|c} 1 & -4 & -1 & -10 \\ 0 & 1 & -2 & -4 \\ 0 & 0 & -25 & -75 \end{array} \right) \Rightarrow_{\frac{1}{-25} R_3 \to R_3} \left(\begin{array}{ccc|c} 1 & -4 & -1 & -10 \\ 0 & 1 & -2 & -4 \\ 0 & 0 & 1 & 3 \end{array} \right) \tag{3.30}$$

$$\left(\begin{array}{ccc|c} 1 & -4 & -1 & -10 \\ 0 & 1 & -2 & -4 \\ 0 & 0 & 1 & 3 \end{array} \right) \Rightarrow_{2R_3 + R_2 \to R_2} \left(\begin{array}{ccc|c} 1 & -4 & -1 & -10 \\ 0 & 1 & 0 & 2 \\ 0 & 0 & 1 & 3 \end{array} \right) \tag{3.31}$$

$$\begin{pmatrix} 1 & -4 & -1 & -10 \\ 0 & 1 & 0 & 2 \\ 0 & 0 & 1 & 3 \end{pmatrix} \Rightarrow_{4R_2 + R_1 \to R_1} \begin{pmatrix} 1 & 0 & -1 & -2 \\ 0 & 1 & 0 & 2 \\ 0 & 0 & 1 & 3 \end{pmatrix} \tag{3.32}$$

$$\begin{pmatrix} 1 & 0 & -1 & -2 \\ 0 & 1 & 0 & 2 \\ 0 & 0 & 1 & 3 \end{pmatrix} \Rightarrow_{R_1 + R_3 \to R_1} \begin{pmatrix} 1 & 0 & 0 & 1 \\ 0 & 1 & 0 & 2 \\ 0 & 0 & 1 & 3 \end{pmatrix} \tag{3.33}$$

Once the left hand side equals the unity matrix, you are done. Let's put it back into the $A\vec{x} = \vec{d}$ notation to make this more obvious:

$$\begin{pmatrix} 1 & 0 & 0 \\ 0 & 1 & 0 \\ 0 & 0 & 1 \end{pmatrix} \begin{pmatrix} x \\ y \\ z \end{pmatrix} = \begin{pmatrix} 1 \\ 2 \\ 3 \end{pmatrix} \tag{3.34}$$

Now multiply out the matrix and you will get back to the linear equation form and see that $x = 1$; $y = 2$; $z = 3$.

3.2.1 Underdetermined Systems

You will sometimes have fewer equations than there are unknowns, or that two or more of the equations are linear combinations of each other. In this case, the system is said to be "underdetermined". That's ok, we can still apply the Gauss elimination method to simplify the system. For example, let us solve the following equations:

$$2x + 2y - 2z = 0 \tag{3.35a}$$
$$x + 3y - z = 0 \tag{3.35b}$$
$$-x + y + z = 0 \tag{3.35c}$$

We'll use the same approach as before:

$$\begin{pmatrix} 2 & 2 & -2 & 0 \\ 1 & 3 & -1 & 0 \\ -1 & 1 & 1 & 0 \end{pmatrix} \to \begin{pmatrix} 1 & 3 & -1 & 0 \\ 0 & 4 & 0 & 0 \\ 1 & 1 & -1 & 0 \end{pmatrix} \to \begin{pmatrix} 1 & 3 & -1 & 0 \\ 0 & 1 & 0 & 0 \\ 0 & 2 & 0 & 0 \end{pmatrix} \to \begin{pmatrix} 1 & 0 & -1 & 0 \\ 0 & 1 & 0 & 0 \\ 0 & 0 & 0 & 0 \end{pmatrix}$$
$$\tag{3.36}$$

Since the last row reduces to all zeroes, all we can say about this set of equations is

$$x = z \tag{3.37}$$
$$y = 0 \tag{3.38}$$

Or, we can *parameterize* the solution by writing

$$x = t \tag{3.39}$$
$$y = 0 \tag{3.40}$$
$$z = t \tag{3.41}$$

where t is an arbitrary parameter.

Homework 3.3: Solving a System of Linear Equations

Solve the following set of equations using matrix row reduction:

$$
\begin{array}{rrrr}
2x+ & 3y- & z = & 3 \\
x+ & & 2z = & -8 \\
4x- & y+ & z = & 5
\end{array}
\tag{3.42}
$$

3.3 DETERMINANTS

What is a determinant? It is an array of numbers that represents a certain value, as given by a well-defined rule set. Determinants are related to arrays and, in fact, are important in their useage.

Suppose we are given a system of two equations:

$$a_{11}x_1 + a_{12}x_2 = d_1 \tag{3.43a}$$
$$a_{21}x_1 + a_{22}x_2 = d_2 \tag{3.43b}$$

Notice that instead of writing the variables as x and y, we have chosen to use the more general notation of x_i. Again, we will put this into matrix multiplication form:

$$A\vec{x} = \vec{d} \tag{3.44}$$

with

$$A = \begin{pmatrix} a_{11} & a_{12} \\ a_{21} & a_{22} \end{pmatrix} \tag{3.45}$$

$$\vec{x} = \begin{pmatrix} x_1 \\ x_2 \end{pmatrix} \tag{3.46}$$

$$\vec{d} = \begin{pmatrix} d_1 \\ d_2 \end{pmatrix} \tag{3.47}$$

One can show that a solution to Eq. 3.44 exists if and only if

$$x_1 = \frac{d_1 a_{22} - d_2 a_{12}}{\Delta_2} \tag{3.48a}$$

$$x_2 = \frac{d_2 a_{11} - d_1 a_{21}}{\Delta_2} \tag{3.48b}$$

where Δ_2 is the *determinant* of A and is defined as

$$\Delta_2 \equiv \det \begin{pmatrix} a_{11} & a_{12} \\ a_{21} & a_{22} \end{pmatrix} \tag{3.49a}$$

$$\equiv \begin{vmatrix} a_{11} & a_{12} \\ a_{21} & a_{22} \end{vmatrix} \tag{3.49b}$$

$$\equiv a_{11}a_{22} - a_{12}a_{21} \tag{3.49c}$$

Similarly, it can be shown that the general 3×3 linear system can be solved if and only if Δ_3 is non-zero, where

$$\Delta_3 \equiv \det \begin{pmatrix} a_{11} & a_{12} & a_{13} \\ a_{21} & a_{22} & a_{23} \\ a_{31} & a_{32} & a_{33} \end{pmatrix} \tag{3.50a}$$

$$\equiv \begin{vmatrix} a_{11} & a_{12} & a_{13} \\ a_{21} & a_{22} & a_{23} \\ a_{31} & a_{32} & a_{33} \end{vmatrix} \tag{3.50b}$$

3.3.1 Computation of Determinants

Determinants of rank 3 or higher can readily be computed using extensions of the following algorithm:

$$\det \begin{pmatrix} a & b & c \\ d & e & f \\ g & h & k \end{pmatrix} = \begin{vmatrix} a & b & c \\ d & e & f \\ g & h & k \end{vmatrix} = a \begin{vmatrix} e & f \\ h & k \end{vmatrix} - b \begin{vmatrix} d & f \\ g & k \end{vmatrix} + c \begin{vmatrix} d & e \\ g & h \end{vmatrix} \tag{3.51a}$$

That is, you can break up a 3×3 determinant into a sum of sub-determinants. Note the alternating sign. Here's a concrete example:

$$\begin{vmatrix} 1 & 2 & 1 \\ 3 & 0 & 4 \\ 8 & 4 & 10 \end{vmatrix} = (1) \begin{vmatrix} 0 & 4 \\ 4 & 10 \end{vmatrix} - (2) \begin{vmatrix} 3 & 4 \\ 8 & 10 \end{vmatrix} + (1) \begin{vmatrix} 3 & 0 \\ 8 & 4 \end{vmatrix} = 0 . \tag{3.52a}$$

Homework 3.4: Computing Determinants

Find the values of the following determinants:

$$\begin{vmatrix} 1 & 2 \\ 3 & 4 \end{vmatrix} , \quad \begin{vmatrix} 1 & -2 \\ 3 & 4 \end{vmatrix} , \quad \begin{vmatrix} 7 & 1 & 9 \\ -2 & 9 & 0 \\ 3 & 1 & 4 \end{vmatrix} .$$

3.4 EIGENVALUES AND EIGENVECTORS

Eigenvalues and eigenvectors are closely tied to *linear transformations*. Consider the following matrix expression:

$$A\vec{x} . \tag{3.53}$$

If

$$A = \begin{pmatrix} 2 & 1 \\ 1 & 2 \end{pmatrix} \tag{3.54}$$

and if $\vec{x} \equiv (x, y)$, then

$$\begin{pmatrix} 2 & 1 \\ 1 & 2 \end{pmatrix} \begin{pmatrix} x \\ y \end{pmatrix} = \begin{pmatrix} 2x + y \\ x + 2y \end{pmatrix} \tag{3.55}$$

This is a linear transformation. For example, the coordinate $(2, 2)$ will be transformed to the new coordinate

$$\begin{pmatrix} 2 & 1 \\ 1 & 2 \end{pmatrix} \begin{pmatrix} 2 \\ 2 \end{pmatrix} = \begin{pmatrix} (2)(2) + (1)(2) \\ (1)(2) + (2)(2) \end{pmatrix} = \begin{pmatrix} 6 \\ 6 \end{pmatrix} \tag{3.56}$$

The coordinate $(1, 2)$ will be transformed to the new coordinate

$$\begin{pmatrix} 2 & 1 \\ 1 & 2 \end{pmatrix} \begin{pmatrix} 1 \\ 2 \end{pmatrix} = \begin{pmatrix} (2)(1) + (1)(2) \\ (1)(1) + (2)(2) \end{pmatrix} = \begin{pmatrix} 4 \\ 5 \end{pmatrix} \tag{3.57}$$

To get a physical idea of a linear transformation, look at Figs. 3.1 and 3.2. In Fig. 3.1 we depict several vectors on the x-y axis space. In Fig. 3.2 we show those same vectors after being linearly transformed by Eq. 3.54. Indeed, we have transformed the entire x-y space, as seen by the distortion of the coordinate system itself. What you are seeing is a linear transformation that looks like a stretching action along one of the diagonal axes. This is the linear transformation corresponding to matrix A above. (We'll prove this soon.) Now look at the red arrow that is in the upper right quadrant of the figure. Before the linear transformation the tail of the arrow is at coordinate $(2, 2)$; after the transformation, the tail's coordinate is at $(6, 5)$. Before and after the transformation, the head of that same arrow is at $(1, 2)$ and $(4, 5)$, respectively. The head and tail of the red arrow are clearly following the same linear transformation as shown in Eq. 3.53.

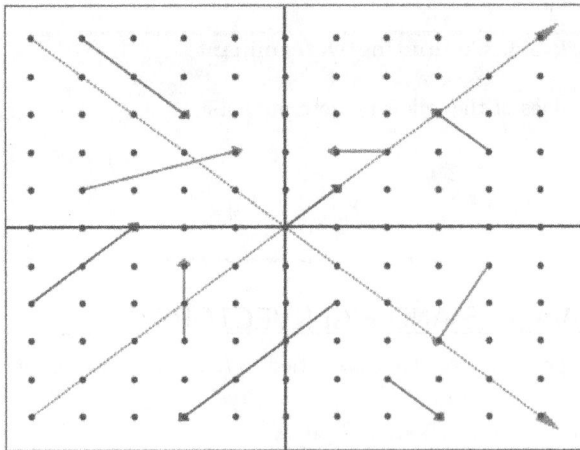

FIGURE 3.1: A plot of several vectors.

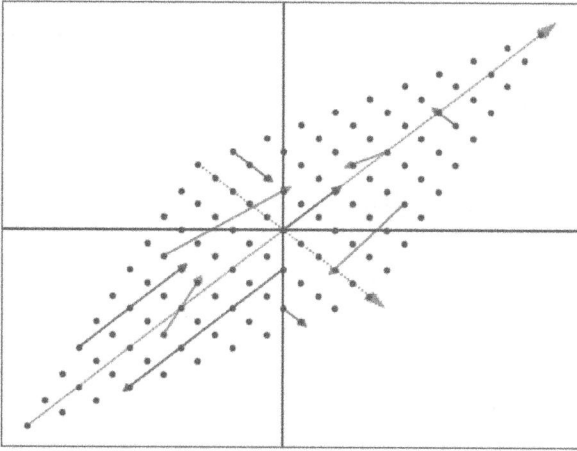

FIGURE 3.2: A plot of the same vectors in Fig. 3.1 but after the linear transformation of Eq. 3.54.

The red arrows are distorted by the transformation in that both their lengths and directions have changed. By comparison, look at the magenta arrows. They change in neither length nor direction. The blue arrows, on the other hand, change in length, but not in direction. The question is, how to characterize these phenomena?

The answer, in short, is through "eigen-analysis". Consider the following matrix equation:

$$A\vec{x} = \lambda\vec{x} \tag{3.58}$$

In Eq. 3.58, applying a linear transformation on the vector $\vec{x}$ gives us back the same vector, but multiplied by some constant, λ. This corresponds to the cases of the magenta arrows, for which $\lambda = 1$, and the blue arrows, for which $\lambda = 3$. Equation 3.58 is called an eigen-equation for which λ is the eigenvalue and $\vec{x}$ is the eigenvector. The equation is telling us that for a particular choice of coordinate system (say, direction of stretch), an arrow will not be distorted, but will merely be stretched (or shrunk) by a factor of λ. Solving the eigenvalue equation will give us the values of all the eigenvalues (one for each dimension of the system, here 2), and the eigenvectors, also one vector for each dimension in the system. **We note that the time-independant Schrödinger equation is an eigenvalue/eigenvector equation, where the matrix A is replaced by a differential operator and the vector $\vec{x}$ is replaced by the wave function.**

This is all very interesting, but how to solve this equation for the λs and $\vec{x}$s? We start by rewriting Eq. 3.58 as

$$(A - \lambda\,1)\,\vec{x} = 0 \tag{3.59}$$

Equation 3.59 will be satisfied if and only if

$$\det(A - \lambda\,1) = 0 \tag{3.60}$$

where 1 is the unity matrix. Therefore, we need only solve Eq. 3.60 to obtain the eigenvalues of A.

3.4.1 Example 1

3.4.1.1 Eigenvalues

For the above value of A, Eq. 3.60 becomes

$$\begin{vmatrix} 2 - \lambda & 1 \\ 1 & 2 - \lambda \end{vmatrix} = (2 - \lambda)^2 - 1 = (\lambda - 3)(\lambda - 1) = 0 \tag{3.61}$$

Equation 3.61 has the roots $\lambda_{1,2} = 3, 1$.

3.4.1.2 Eigenvectors

One by one, we can plug these values of λ into Eq. 3.59 and solve for the eigenvectors. We will call the eigenvector that corresponds to λ_n u_n. Then, for $\lambda_1 = 3$,

$$\left(\begin{array}{cc|c} (2-3) & 1 & 0 \\ 1 & (2-3) & 0 \end{array} \right) \Rightarrow \left(\begin{array}{cc|c} -1 & 1 & 0 \\ 1 & -1 & 0 \end{array} \right) \Rightarrow \left(\begin{array}{cc|c} 1 & -1 & 0 \\ 0 & 0 & 0 \end{array} \right) \tag{3.62}$$

Thus, $x_1 = x_2 = t$. Arbitrarily choosing $t = 1$ gives us

$$u_1 = \begin{pmatrix} 1 \\ 1 \end{pmatrix} \tag{3.63}$$

Note that we only know the eigenvectors to within an arbitrary constant. This is clear from Eq. 3.58: we can multiply both sides of that equation by a constant, absorbing the constant into $\vec{x}$, without changing the equation.

Following the same procedure as in Eq. 3.62, but for $\lambda_2 = 1$, gives us

$$u_2 = \begin{pmatrix} -1 \\ 1 \end{pmatrix} \tag{3.64}$$

Before moving on to another example, let's look at what we've just done. As we can see from Figs. 3.1 and 3.2, any arrow that lies entirely along *either* u_1 or u_2 will be stretched by an amount equal to the corresponding eigenvalue, but its orientation will not change. Now, any arrow can be drawn to have its components along a combination of u_1 and u_2. But because the eigenvalues corresponding to those eigenvectors are not in general equal, the arrow's direction will change.

In the case of the above example, the eigenvector $u_1 = (1, 1)$, corresponding to the eigenvalue $\lambda_1 = 3$ means that if we stretch along the "1-1 axis", that is at 45 degrees counter-clockwise from the $+x$ axis, any arrow that is parallel to u_1 will be stretched by a factor of 3 – which is what we observe. For $u_2 = (-1, 1)$, corresponding to the eigenvalue $\lambda_2 = 1$, we see that if we stretch along the "1, -1 axis", that is at 45 degrees clockwise from the $= x$ axis, any arrow that is parrallel to u_2 will be stretched by a factor of 1, that is to say, that it will retain its original length. This is also what we observe. So we see that the eigenvalue/eigenvector equation tells us something very special about the matrix it contains: it tells us about the "special" axes associated with that particular linear transformation.

3.4.1.3 Normalization

Often we wish our eigenvectors to be *normalized*. That is, we want

$$u^T u = 1 \qquad (3.65)$$

We can always multiply our eigenvectors by some constant to achieve this normalization. In the above example, we stopped short of the normalization process; in Example 2, we normalize our eigenvectors.

3.4.2 Example 2

3.4.2.1 Eigenvalues

Consider the matrix

$$A = \begin{pmatrix} 2 & -4 \\ -1 & -1 \end{pmatrix} \qquad (3.66)$$

Then the equation we need to solve is

$$\begin{pmatrix} 2 & -4 \\ -1 & -1 \end{pmatrix} \begin{pmatrix} x_1 \\ x_2 \end{pmatrix} = \lambda \begin{pmatrix} x_1 \\ x_2 \end{pmatrix} \qquad (3.67)$$

Rearranging Eq. 3.67, we get

$$\begin{pmatrix} 2 - \lambda & -4 \\ -1 & -1 - \lambda \end{pmatrix} \begin{pmatrix} x_1 \\ x_2 \end{pmatrix} = \begin{pmatrix} 0 \\ 0 \end{pmatrix} \qquad (3.68)$$

Equation 3.68 is true if and only if the determinant of matrix $(A - \lambda 1)$ is zero. That is,

$$\begin{vmatrix} 2 - \lambda & -4 \\ -1 & -1 - \lambda \end{vmatrix} = 0 \qquad (3.69)$$

Then,

$$(2 - \lambda)(-1 - \lambda) - (-4)(-1) = 0 \qquad (3.70a)$$
$$\lambda^2 - \lambda - 6 = 0 \qquad (3.70b)$$
$$(\lambda - 3)(\lambda + 2) = 0 \qquad (3.70c)$$

The two eigenvalues are therefore

$$\lambda_1 = -2 \qquad (3.71a)$$
$$\lambda_2 = 3 \qquad (3.71b)$$

3.4.2.2 Eigenvectors

Now that we have the eigenvalues, we can use Eq. 3.59 to solve for the eigenvectors, one for each eigenvalue. Starting with $\lambda_1 = -2$, we get the matrix equation

$$\begin{pmatrix} 2-(-2) & -4 & \Big| & 0 \\ -1 & -1-(-2) & \Big| & 0 \end{pmatrix} = \begin{pmatrix} 4 & -4 & \Big| & 0 \\ -1 & 1 & \Big| & 0 \end{pmatrix} = \begin{pmatrix} 1 & -1 & \Big| & 0 \\ 0 & 0 & \Big| & 0 \end{pmatrix} \quad (3.72)$$

Equation 3.72 is under-determined. But we can say that

$$x_1 = x_2 \quad (3.73a)$$
$$x_2 = x_2 \quad (3.73b)$$

Therefore,

$$u_1 = k \begin{pmatrix} 1 \\ 1 \end{pmatrix} \quad (3.74)$$

Now, if we wish our vector to be a unit vector, then $|u_1|^2 = 1$. That is,

$$k^2 \left((1)^2 + (1)^2 \right) = 1 \quad (3.75)$$

Or, $k = 1/\sqrt{2}$. Then,

$$u_1 = \frac{1}{\sqrt{2}} \begin{pmatrix} 1 \\ 1 \end{pmatrix} \quad (3.76)$$

Now we repeat the process for $\lambda_2 = 3$.

$$\begin{pmatrix} 2-3 & -4 & \Big| & 0 \\ -1 & -1-3 & \Big| & 0 \end{pmatrix} = \begin{pmatrix} -1 & -4 & \Big| & 0 \\ -1 & -4 & \Big| & 0 \end{pmatrix} = \begin{pmatrix} -1 & -4 & \Big| & 0 \\ 0 & -4 & \Big| & 0 \end{pmatrix} \quad (3.77)$$

Equation 3.77 is under-determined. But we can say that

$$x_1 = -4x_2 \quad (3.78a)$$
$$x_2 = x_2 \quad (3.78b)$$

Therefore,

$$u_2 = k \begin{pmatrix} -4 \\ 1 \end{pmatrix} \quad (3.79)$$

Now, if we wish our vector to be a unit vector, then $|u_2|^2 = 1$. That is,

$$k^2 \left((-4)^2 + (1)^2 \right) = 1 \quad (3.80)$$

Or, $k = 1/\sqrt{17}$. Then,

$$u_2 = \frac{1}{\sqrt{17}} \begin{pmatrix} -4 \\ 1 \end{pmatrix} \quad (3.81)$$

So the two eigenvalue/eigenvector combinations for matrix A are

$$\lambda_1 = -2, \; u_1 = \frac{1}{\sqrt{2}} \begin{pmatrix} 1 \\ 1 \end{pmatrix} \quad (3.82a)$$

$$\lambda_2 = 3, \; u_2 = \frac{1}{\sqrt{17}} \begin{pmatrix} -4 \\ 1 \end{pmatrix} \quad (3.82b)$$

3.4.3 Degenerate Eigenvalues

It is possible that, upon solving 3.60 for the eigenvalues, that they are not unique. In this case we say two or more of the eigenvalues are *degenerate*. But when this happens, how do you come up with independent eigenvectors? Simple: you make use of the ortho-normality of eigenvectors. This can be expressed as

$$\vec{u}_n \cdot \vec{u}_m = \delta_{nm} \tag{3.83}$$

Here, δ_{nm} is the so-called Kronecker delta function. Its value is 1 for $n = m$ and 0 otherwise. Let's see how this works with an example.

3.4.3.1 Example of Degenerate Eigenvalues

Consider the matrix

$$A = \begin{pmatrix} 4 & 0 & -6 \\ -3 & -2 & 3 \\ 3 & 0 & -5 \end{pmatrix} . \tag{3.84}$$

The determinant equation we need to solve in order to obtain the eigenvalues of A is

$$\begin{vmatrix} 4 - \lambda & 0 & -6 \\ -3 & -2 - \lambda & 3 \\ 3 & 0 & -5 - \lambda \end{vmatrix} = 0 . \tag{3.85}$$

You can show (see the homework!) that this leads to the eigenvalues,

$$\lambda_1 = 1 \tag{3.86a}$$
$$\lambda_2 = -2 \tag{3.86b}$$
$$\lambda_3 = -2 \tag{3.86c}$$

Inserting λ_1 into Eq. 3.60 and solving for the corresponding eigenvector, $\vec{u}_1$, gives us (after normalization):

$$\vec{u}_1 = \frac{1}{\sqrt{6}} (2, -1, 1) . \tag{3.87}$$

However, when we try this same procedure with λ_2, we obtain

$$\left(\begin{array}{ccc|c} 1 & 0 & -1 & 0 \\ 0 & 0 & 0 & 0 \\ 0 & 0 & 0 & 0 \end{array} \right) . \tag{3.88}$$

In other words, we could say that

$$\vec{u}_2 = k_2 (1, y, 1) \tag{3.89}$$

where k_2 is the normalization constant. That is, we can parameterize the under-determined equations with $z = 1$, but that still leaves y unknown. However, we can now make use of Eq. 3.83:

$$\vec{u}_1 \cdot \vec{u}_2 = 0 \rightarrow (2, -1, 1) \cdot (1, y, 1) = 0 . \tag{3.90}$$

Or,

$$(2)(1) + (-1)(y) + (1)(1) = 0 , \tag{3.91}$$

which gives us

$$y = 3. \tag{3.92}$$

Therefore, after normalization,

$$\vec{u}_2 = \frac{1}{\sqrt{11}} \, (1, 3, 1) . \tag{3.93}$$

Now to obtain $\vec{u}_3$ we use Eq. 3.83 twice:

$$\vec{u}_1 \cdot \vec{u}_3 = 0 \tag{3.94a}$$
$$\vec{u}_2 \cdot \vec{u}_3 = 0 , \tag{3.94b}$$

which, after normalization gives, along with the other eigenvectors:

$$\vec{u}_1 = \frac{1}{\sqrt{6}} \, (2, -1, 1) \tag{3.95a}$$

$$\vec{u}_2 = \frac{1}{\sqrt{11}} \, (1, 3, 1) \tag{3.95b}$$

$$\vec{u}_3 = \frac{1}{\sqrt{66}} \, (-4, -1, 7) \tag{3.95c}$$

Homework 3.5: Computing Eigenvectors Having Degenerate Eigenvalues

Starting from Eq. 3.84, fill in the missing parts to obtain the eigenvalues of Eqs. 3.86 and eigenvectors of Eqs. 3.95.

3.5 DIAGONALIZING A MATRIX

Once you've computed the eigenvalues and *normalized* eigenvectors of a matrix, you can *diagonalize* that matrix, placing the eigenvalues along the main diagonal. The diagonalized matrix is simply

$$\begin{pmatrix} \lambda_1 & 0 & 0 & \cdots & 0 \\ 0 & \lambda_2 & & & \\ 0 & 0 & \ddots & & \\ \vdots & & & & \\ 0 & & & & \lambda_n \end{pmatrix} = P^{-1}AP \tag{3.96}$$

where P is the array made up of columns of *normalized* eigenvectors.

For the matrix of Example 1 above,

$$A = \begin{pmatrix} 2 & 1 \\ 1 & 2 \end{pmatrix} , P = \frac{1}{\sqrt{2}} \begin{pmatrix} 1 & -1 \\ 1 & 1 \end{pmatrix} \tag{3.97}$$

where P is the matrix of eigenvectors, from above. Then

$$P^{-1} = \frac{1}{\sqrt{2}} \begin{pmatrix} 1 & 1 \\ -1 & 1 \end{pmatrix} \tag{3.98}$$

Using Eq. 3.96 with these values of A, P, and P^{-1} we obtain:

$$P^{-1}AP = \frac{1}{2} \begin{pmatrix} 1 & -1 \\ 1 & 1 \end{pmatrix} \begin{pmatrix} 2 & 1 \\ 1 & 2 \end{pmatrix} \begin{pmatrix} 1 & 1 \\ -1 & 1 \end{pmatrix} = \begin{pmatrix} 3 & 0 \\ 0 & 1 \end{pmatrix} \tag{3.99}$$

which is a diagonal matrix with the eigenvalues along the diagonal.

3.6 COMPLEX MATRICES

Thus far, all of our examples involved matrices having pure real elements, but this need not be the case. Consider

$$A = \begin{pmatrix} 0 & i \\ -i & 0 \end{pmatrix} \tag{3.100}$$

We find the eigenvalues in the usual way:

$$\begin{vmatrix} 0 - \lambda & i \\ -i & 0 - \lambda \end{vmatrix} = 0 \tag{3.101}$$

Or,

$$\lambda^2 - 1 = 0 . \tag{3.102}$$

from which we obtain

$$\lambda_1 = +1 \tag{3.103a}$$
$$\lambda_2 = -1 \tag{3.103b}$$

We can now determine the eigenvectors in the usual way. For $\lambda_1 = +1$:

$$\begin{pmatrix} -1 & i & | & 0 \\ -i & -1 & | & 0 \end{pmatrix} \Rightarrow \begin{pmatrix} 1 & -i & | & 0 \\ i & -i & | & 0 \end{pmatrix} \Rightarrow \begin{pmatrix} 1 & -i & | & 0 \\ 0 & 0 & | & 0 \end{pmatrix} \tag{3.104}$$

Therefore,

$$\mathbf{u_1} = k_1 \begin{pmatrix} i \\ 1 \end{pmatrix} = \frac{1}{\sqrt{2}} \begin{pmatrix} i \\ 1 \end{pmatrix} \tag{3.105}$$

For $\lambda_2 = -1$:

$$\begin{pmatrix} 1 & i & | & 0 \\ -i & 1 & | & 0 \end{pmatrix} \Rightarrow \begin{pmatrix} 1 & i & | & 0 \\ i & i & | & 0 \end{pmatrix} \Rightarrow \begin{pmatrix} 1 & i & | & 0 \\ 0 & 0 & | & 0 \end{pmatrix} \tag{3.106}$$

Therefore,

$$\mathbf{u_2} = k_2 \begin{pmatrix} -i \\ 1 \end{pmatrix} = \frac{1}{\sqrt{2}} \begin{pmatrix} -i \\ 1 \end{pmatrix} \tag{3.107}$$

Wait! Let's back up a step. When I normalized $\mathbf{u_1}$ and $\mathbf{u_2}$ I didn't simply sum the squares of their components. Remember, you can think of a dot product of two vectors $\mathbf{u_1}$ and $\mathbf{u_2}$ as

$$\mathbf{u_1} \cdot \mathbf{u_2} = \overline{\mathbf{u}}_1^{\mathbf{T}} \mathbf{u_2} = \frac{1}{\sqrt{2}} \frac{1}{\sqrt{2}} \begin{pmatrix} -i & 1 \end{pmatrix} \begin{pmatrix} -i \\ 1 \end{pmatrix} = 0$$

where I have pulled the two normalizing factors off to the left and I am using the "overhead bar" to indicate the complex conjugate of all of the elements in the matrix. The superscript T, as before, indicates transpose.

We see then, as expected, that the product of the two different eigenvectors is zero. *This would not have been the case if we had not taken the complex conjugate!* Furthermore, if we look at the inner product (dot product) of a vector with itself, we must use the "transpose-conjugate" there as well:

$$\mathbf{u_1} \cdot \mathbf{u_1} = \overline{\mathbf{u}}_1^{\mathbf{T}} \mathbf{u_1} = \frac{1}{\sqrt{2}} \frac{1}{\sqrt{2}} \begin{pmatrix} -i & 1 \end{pmatrix} \begin{pmatrix} i \\ 1 \end{pmatrix} = 1$$

The point is that *in general* in multiplying two vectors $\mathbf{a}$ and $\mathbf{b}$, we must take the complex transpose of the left vector, and multiply the right vector by that. We just didn't mention the complex conjugate part before because all of the elements in our examples up until now were pure real. This is not unreasonable: after all, the inner product of a vector with itself should result in the length or magnitude of the vector. And as we learned in Chapter 2, one obtains the magnitude of an expression by multiplying it by its complex conjugate and taking the square root.

Note that pure real matrices can have complex eigenvalues and eigenvectors. In the particular example above, our matrix was complex, but a special kind of complex: that matrix was *Hermitian.* A matrix is said to be Hermitian if

$$a_{ij} = a_{ji}^{*}$$

That is, upon reflection about the matrix's diagonal, the lower triangle of elements are the complex conjugates of their counterparts on the upper triangle of elements. Note that, by definition, the diagonal of a Hermitian matrix must be pure real. Furthermore, we can show that the eigenvalues of a Hermitian matrix *will always be pure real.*

Hermitian matrices play an important role in quantum mechanics. Because eigenvalues in quantum mechanics represent observable quantities, they must always be pure real. Therefore, quantum mechanical matrix operators *must* be Hermitian.

Note, however, that the eigenvectors need not be pure real – indeed, they rarely are if the matrix contains complex elements. Therefore it is particularly important to remember to take the *complex* transpose when multiplying vectors in quantum mechanics.

Here's a more challenging example of a Hermitian matrix, from which we wish to deduce the eigenvalues and eigenvectors.

Homework 3.6: Using Complex Matrices

$$A = \begin{pmatrix} \frac{1}{2} & \frac{\sqrt{6}}{6} & \frac{\sqrt{3}i}{6} \\ \frac{\sqrt{6}}{6} & 0 & \frac{3\sqrt{2}i}{6} \\ \frac{-\sqrt{3}i}{6} & \frac{-3\sqrt{2}i}{6} & \frac{-1}{2} \end{pmatrix}$$

(3.108)

As you can see, matrix A defined in Eq. 3.108 is Hermitian. We therefore expect that the eigenvalues will be pure real.

– Determine the eigenvalues of matrix A.

– Determine the eigenvectors of matrix A

– Normalize the eigenvectors.

– Verify that your 3 eigenvectors are orthonormal.

– Use the eigenvectors to form the transformation matrix P and P^{-1}, as in Eqs. 3.97 and 3.98 in Section 3.5 above.

– Verify that the product $P^{-1}AP$ gives a diagonal matrix with the eigenvalues along the diagonal.

3.7 THE ROTATION MATRIX

Consider rotating a vector in the x-y plane through an angle θ. For example, consider the vector $v = 3i + 4j$, where i and j are the unit vectors in the x and y directions, respectively. The new vector will then be given by

$$v' = Rv$$

where v' is the rotated vector and R is the *rotation matrix* given by

$$R = \begin{pmatrix} \cos\theta & -\sin\theta \\ \sin\theta & \cos\theta \end{pmatrix}$$

Then,

$$v' = \begin{pmatrix} \cos\theta & -\sin\theta \\ \sin\theta & \cos\theta \end{pmatrix} \begin{pmatrix} 3 \\ 4 \end{pmatrix} = \begin{pmatrix} 3\cos\theta - 4\sin\theta \\ 3\sin\theta + 4\cos\theta \end{pmatrix}$$

(Note that this rotation did *not* change the length of the vector.)

For the specific case of $\theta = 45°$, we get

$$v' = \begin{pmatrix} \frac{-\sqrt{2}}{2} \\ \frac{7\sqrt{2}}{2} \end{pmatrix} = \frac{-\sqrt{2}}{2}i + \frac{7\sqrt{2}}{2}j$$

In three dimensions, the rotation matrix can be expressed as

$$R = \begin{pmatrix} \cos(x',x) & \cos(y',x) & \cos(z',x) \\ \cos(x',y) & \cos(y',y) & \cos(z',y) \\ \cos(x',z) & \cos(y',z) & \cos(z',z) \end{pmatrix}$$

where, for example, x',x is the angle between the x' and x axes.

We can see how the 3-D rotation matrix reduces to the 2-D case by considering rotation about the z-axis. In this case, the angle between z' and z stays 0, and the angles between x' and z; y' and z; z' and x; and z' and y remain 90°. Then,

$$R = \begin{pmatrix} \cos\theta & -\sin\theta & 0 \\ \sin\theta & \cos\theta & 0 \\ 0 & 0 & 1 \end{pmatrix}$$

So what about the eigenvalues and eigenvectors of R? We'll look at the 2-D case for simplicity, but the same approach can be applied to the 3-D rotational matrix.

We want to solve the equation

$$Rv = \lambda v$$

Following the usual process,

$$|R - \lambda 1| = 0$$

Or,

$$\begin{vmatrix} \cos(\theta) - \lambda & -\sin(\theta) \\ \sin(\theta) & \cos\theta - \lambda \end{vmatrix} = 0 \Rightarrow [\cos\theta - \lambda][\cos\theta - \lambda] + \sin^2\theta = 0$$

giving

$$\lambda_1 = e^{-i\theta} \tag{3.109a}$$

$$\lambda_2 = e^{i\theta} \tag{3.109b}$$

Now solving for the eigenvector corresponding to λ_1,

$$\begin{pmatrix} \cos\theta - e^{-i\theta} & -\sin\theta & \bigg| & 0 \\ \sin\theta & \cos\theta - e^{-i\theta} & \bigg| & 0 \end{pmatrix} \Rightarrow \frac{1}{2}\begin{pmatrix} e^{i\theta} + e^{-i\theta} - 2e^{-i\theta} & +i\left(e^{i\theta} - e^{-i\theta}\right) & \bigg| & 0 \\ -i\left(e^{i\theta} - e^{-i\theta}\right) & e^{i\theta} + e^{-i\theta} - 2e^{-i\theta} & \bigg| & 0 \end{pmatrix} \Rightarrow$$

$$\frac{1}{2}\begin{pmatrix} e^{i\theta} - e^{-i\theta} & +ie^{i\theta} - ie^{-i\theta} & \bigg| & 0 \\ -ie^{i\theta} + ie^{-i\theta} & e^{i\theta} - e^{-i\theta} & \bigg| & 0 \end{pmatrix} \Rightarrow \frac{1}{2}\begin{pmatrix} 2i\sin\theta & -2\sin\theta & \bigg| & 0 \\ 2\sin\theta & 2i\sin\theta & \bigg| & 0 \end{pmatrix} \Rightarrow \frac{1}{2}\begin{pmatrix} i & -1 & \bigg| & 0 \\ 0 & 0 & \bigg| & 0 \end{pmatrix}$$

So,

$$u_1 = k_1\begin{pmatrix} -i \\ 1 \end{pmatrix} = \frac{1}{\sqrt{2}}\begin{pmatrix} -i \\ 1 \end{pmatrix}$$

Now, for $\lambda_2 = e^{i\theta}$,

$$\begin{pmatrix} \cos\theta - e^{i\theta} & -\sin\theta & \bigg| & 0 \\ \sin\theta & \cos\theta - e^{i\theta} & \bigg| & 0 \end{pmatrix} \Rightarrow \begin{pmatrix} e^{i\theta} + e^{-i\theta} - 2e^{i\theta} & -2\sin\theta & \bigg| & 0 \\ 2\sin\theta & e^{i\theta} + e^{-i\theta} - 2e^{i\theta} & \bigg| & 0 \end{pmatrix} \Rightarrow$$

$$\begin{pmatrix} -e^{i\theta} + e^{-i\theta} & -2\sin\theta & \bigg| & 0 \\ 2\sin\theta & -e^{i\theta} + e^{-i\theta} & \bigg| & 0 \end{pmatrix} \Rightarrow \begin{pmatrix} -i\sin\theta & -\sin\theta & \bigg| & 0 \\ \sin\theta & -i\sin\theta & \bigg| & 0 \end{pmatrix} \Rightarrow \begin{pmatrix} 1 & -1 & \bigg| & 0 \\ 0 & 0 & \bigg| & 0 \end{pmatrix}$$

So,

$$u_2 = k_2\begin{pmatrix} i \\ 1 \end{pmatrix} = \frac{1}{\sqrt{2}}\begin{pmatrix} i \\ 1 \end{pmatrix}$$

Now, just for fun, let's diagonalize the 2-D rotation matrix. Recall that

$$P^{-1}AP = \Lambda 1$$

where P is the matrix whose columns are the normalized eigenvectors of A, and $\Lambda 1$ is the diagonal matrix whose elements are the eigenvalues of A. Then,

$$P = \frac{1}{\sqrt{2}} \begin{pmatrix} -i & i \\ 1 & 1 \end{pmatrix}$$

We now have to find P^{-1}, which we do in the usual way:

$$\left(\begin{array}{cc|cc} -i & i & 1 & 0 \\ 1 & 1 & 0 & 1 \end{array} \right) \Rightarrow \left(\begin{array}{cc|cc} 1 & -1 & i & 0 \\ 1 & 1 & 0 & 1 \end{array} \right) \Rightarrow \left(\begin{array}{cc|cc} 1 & -1 & i & 0 \\ 0 & -2 & i & -1 \end{array} \right) \Rightarrow$$

$$\left(\begin{array}{cc|cc} 1 & -1 & i & 0 \\ 0 & 1 & \frac{-i}{2} & \frac{1}{2} \end{array} \right) \Rightarrow \left(\begin{array}{cc|cc} 1 & 0 & \frac{i}{2} & \frac{1}{2} \\ 0 & 1 & \frac{-i}{2} & \frac{1}{2} \end{array} \right) \Rightarrow P^{-1} = \frac{\sqrt{2}}{2} \begin{pmatrix} i & 1 \\ -i & 1 \end{pmatrix}$$

Now we multiply the matrices as indicated above:

$$P^{-1}RP = \frac{1}{2} \begin{pmatrix} i & 1 \\ -i & 1 \end{pmatrix} \begin{pmatrix} \cos\theta & -\sin\theta \\ \sin\theta & \cos\theta \end{pmatrix} \begin{pmatrix} -i & i \\ 1 & 1 \end{pmatrix} \tag{3.110a}$$

$$= \frac{1}{2} \begin{pmatrix} i\cos\theta + \sin\theta & -i\sin\theta + \cos\theta \\ -i\cos\theta + \sin\theta & i\sin\theta + \cos\theta \end{pmatrix} \begin{pmatrix} -i & i \\ 1 & 1 \end{pmatrix} \tag{3.110b}$$

$$= \frac{1}{2} \begin{pmatrix} -i(i\cos\theta + \sin\theta) + (-i\sin\theta + \cos\theta) & i(i\cos\theta + \sin\theta) + (-i\sin\theta + \cos\theta) \\ -i(-i\cos\theta + \sin\theta) + (i\sin\theta + \cos\theta) & i(-i\cos\theta + \sin\theta) + (i\sin\theta + \cos\theta) \end{pmatrix} \tag{3.110c}$$

$$= \frac{1}{2} \begin{pmatrix} 2\cos\theta - 2i\sin\theta & 0 \\ 0 & 2\cos\theta + 2i\sin\theta \end{pmatrix} \tag{3.110d}$$

$$= \begin{pmatrix} e^{-i\theta} & 0 \\ 0 & e^{i\theta} \end{pmatrix} \tag{3.110e}$$

3.8 MATRICES IN PYTHON

We've learned quite a bit about matrices. One of the things we've learned is that it can be tedious to manipulate them by hand! Fortunately, because of their compact and organized structures, matrices lend themselves to manipulation by computer code, obviating the need for tedious manual treatments. Let's start this section by introducing numpy's numerical treatment of matrices. First of all, in numpy matrices are referred as *arrays*. (Note there is also a data type called a *matrix*. The two data types have similar capabilities, but they are *not* interchangeable. In this text, we will stick to the *array* data type.)

3.8.1 Creating Arrays in Python

Let's see how to create an array. One way is to use the **zeros** method. This creates an array that is pre-filled with 0 for each of its elements. The listing shows how to do this with a 1-D array, and the two ways to do this with an $n \times m$ array.

```
1  ' ' '
2  Creating arrays using the zeros method
3  ' ' '
4  import numpy as np
5  a = np.zeros(4)
6  print(a,"\n")
7
8  b = np.zeros((3,5))
9  print(b, "\n")
10
11 c = np.zeros(15)
12 d = np.reshape(c,(3,5))
13 print(d, "\n")
```

Listing 3.1: Sample code showing how to create an array using the **zeros** method.

When I execute this code, I get
[0. 0. 0. 0.]

[[0. 0. 0. 0. 0.]
 [0. 0. 0. 0. 0.]
 [0. 0. 0. 0. 0.]]

[[0. 0. 0. 0. 0.]
 [0. 0. 0. 0. 0.]
 [0. 0. 0. 0. 0.]]

In lines 5, 8, and 11, I used the **zeros** method. Notice that in all cases, the **zeros** method has just a single argument. In lines 5 and 11 I wanted a 1-D array and so the argument was just the desired array length. In line 8, I wanted an array of dimensions 3×5 and so my *sole* argument was a tuple, which contains the desired array dimensions. There is another method, **ones**, which functions just like **zeros** except that it fills the array with 1's instead of 0's.

Sometimes you may find it more convenient to create a 1-D array, and then convert it to a 2-D array, in this case another 3×5 array. Lines 11-13 show a way to do this. First, I used the **zeros** method to create a 1-D array. I then used the **reshape** method to convert this 15 element 1-D array into a 15 element 3×5 array. The **reshape** method has 2 arguments. The first is the name of the array you want to reshape, and the second is a tuple containing the desired dimensions. If the total number of elements in the re-shaped array do not equal the number of elements in the original array, Python will return an error. Another way to use the **reshape** method is:
d = c.reshape(3,2) Note that because c is already of type **array**, we do not need to tell Python to look in **numpy** to find the **reshape** method.

This is all fine and good but, generally speaking, you will like to work with arrays that are more interesting than these all-zero or all-one examples. Let's examine a way to fill an array with something more interesting than the above arrays. The first approach is to create an empty array, and then fill it with values of interest:

```
1  '''
2  Creating an empty array and then filling it using a for-loop
3  '''
4  import numpy as np
5  a = np.empty(0)
6  print(a)
7
8  for i in range(6):
9      a = np.append(a,2*i-1)
10 print(a)
```

Listing 3.2: Sample code showing how to create an empty array, and then fill it with values using a for-loop.

When I execute this code, I get

[]
[-1. 1. 3. 5. 7. 9.]

Of course one way to fill the array with your desired values would be to do so "manually", for example by writing something like,

```
1  '''
2  Filling an array 'manually'
3  '''
4  import numpy as np
5  a = np.ones((2,2))
6  a[0][1] = 4
7  a[1][0] = 2
8  a[1][1] = 2
9  print(a)
```

Listing 3.3: Sample code showing how to manually fill an array.

When I execute this code, I get
[[1. 4.]
 [2. 2.]]

```
1  '''
2  Converting a list to an array
3  '''
4  import numpy as np
5  a_list = [] # create an empty list
6  for i in range(5):
7      a_list.append(i)
8  print("list: ", a_list)
9  a_array = np.array(a_list)
10 print("array: ", a_array)
```

Listing 3.4: Sample code showing how to create a list and then convert it into an array.

When I execute this code, I get
list: [0, 1, 2, 3, 4]
array: [0 1 2 3 4]

As you can see, when printed out, a list and an array appear different only on a superficial level. However, the ways Python uses them are very far apart! Let's start working through the examples given earlier in this chapter, only this time we will use Python to extract our results numerically. First comes multiplication by a scaler.

3.8.2 Multiplying an Array by a Scaler

Using the same matrix (array) as in Eq. 3.3, we can multiply A by 5, where we have already defined array A in Listing 3.3:

```
'''
Multiplying a previously defined array by a scaler
'''
b = 5*a
print(a, "\n")
print(b, "\n")
```

Listing 3.5: Sample code showing how to mulitiply an array by a scaler.

When I execute this code, I get
[[1. 4.]
[2. 2.]]

[[5. 20.]
[10. 10.]]

3.8.3 Array Addition and Multiplication

Well, that was simple enough! Now how about matrix (array) addition and multiplication? Not too bad:

```
'''
Adding and multiplying arrays
'''
import numpy as np
a_for_adding = np.ones((2,3))
a_for_adding[0][1] = 3
a_for_adding[0][2] = -5
a_for_adding[1][0] = 0
a_for_adding[1][1] = 2
a_for_adding[1][2] = 6

b_for_adding = np.ones((2,3))
b_for_adding[0][0] = -3
b_for_adding[0][1] = 2
```

```
15 b_for_adding [0][2] = 4
16 b_for_adding [1][0] = 6
17 b_for_adding [1][1] = −1
18 b_for_adding [1][2] = 7
19
20 a_for_mult = a_for_adding
21
22 b_for_mult = np.ones((3,2))
23 b_for_mult [0][0] = −3
24 b_for_mult [0][1] = 6
25 b_for_mult [1][0] = 2
26 b_for_mult [1][1] = −1
27 b_for_mult [2][0] = 4
28 b_for_mult [2][1] = 7
29
30 print("A (adding)=\n  ", a_for_adding,"\n")
31 print("B (adding)=\n  ", b_for_adding,"\n")
32 print("A + B =\n  ",a_for_adding + b_for_adding, "\n")
33
34 print("A (mult)= ", a_for_mult,"\n")
35 print("B (mult)= ", b_for_mult,"\n")
36 print("A x B =\n  ",a_for_mult @ b_for_mult)
```

Listing 3.6: Sample code showing how to add, and then multiply, two arrays.

When I execute this code, I get

A (adding)=
[[1. 3. -5.]
 [0. 2. 6.]]

B (adding)=
[[-3. 2. 4.]
 [6. -1. 7.]]

A + B =
[[-2. 5. -1.]
 [6. 1. 13.]]

A (mult)=
[[1. 3. -5.]
 [0. 2. 6.]]

B (mult)=
[[-3. 6.]
 [2. -1.]
 [4. 7.]]

A x B =
[[-17. -32.]
 [28. 40.]]

A few things to notice here: While it seems as if I had to write many lines of code just to add and then multiply two arrays, in fact, almost the entire program was occupied with first defining the arrays and printing out the arrays and the results. It took just one line (32) to do the addition, and one more line (36) to do the multiplication. By the way, notice that the addition operator is "+", just as you are used to. Multiplication, on the other hand, uses the "@" symbol because "*" is already reserved for scaler multiplication. Operationally, you will almost always use your programs to compute the values of the elements of your arrays, after which multiplication and addition are trivial.

3.8.4 Numerical Computation of the Inverse of a Matrix (Array)

As an example of using Python to compute the inverse of a matrix, we'll use the array of Eq. 3.15 for the array we wish to invert, and then simply compute its inverse.

```
1  '''
2  Computing the inverse of an array
3  '''
4  import numpy as np
5  A = np.array([[5, 2],[3, 1]])
6  A_inv = np.linalg.inv(A)
7
8  print("A = \n", A, "\n")
9  print("A inverse = \n", A_inv, "\n")
10 print("A x A_inv = \n", A @ A_inv, "\n")  # just to check...
11
12 # That was fun!  Let's do it again for a 3x3 array:
13 B = np.array([[6, 1, 1],
14              [4, -2, 5],
15              [2, 8, 7]])
16 B_inv = np.linalg.inv(B)
17 print("B = \n", B, "\n")
18 print("B inverse = \n", B_inv, "\n")
19 print("B x B_inv = \n", B @ B_inv, "\n")  #again, just to check...
```

Listing 3.7: Sample code showing how to take the inverse of an array.

When I execute this code, I get

$A =$
[[5 2]
[3 1]]

$A^{-1} =$
[[-1. 2.]
[3. -5.]]

$A \times A^{-1} =$
[[1.0000000e+00 0.0000000e+00]
[-4.4408921e-16 1.0000000e+00]]

$B =$
[[6 1 1]
[4 -2 5]
[2 8 7]]

B^{-1} = [[0.17647059 -0.00326797 -0.02287582] [0.05882353 -0.13071895 0.08496732] [-0.11764706 0.1503268 0.05228758]]

$B \times B^{-1}$ = [[1.00000000e+00 0.00000000e+00 6.93889390e-18] [2.77555756e-17 1.00000000e+00 4.85722573e-17] [8.32667268e-17 1.11022302e-16 1.00000000e+00]]

As you can see, this was just as easy for a 3×3 array as it was for a 2×2 array. As a check, I multiplied the original arrays, A and B, by their computed inverses. We should get the unity matrix, but we didn't. What's going on here? Well, the off-diagonal elements in the products of $A \times A^{-1}$ and $B \times B^{-1}$ are due to the inevitable round-off error. If this bothers you (and I think it should because it is difficult to read) refer back to Chapter 1 and remind yourself how to select how many digits precision you'd like to show in your output.

3.8.5 Using Python to Numerically Solve Systems of Coupled Linear Equations

Earlier, we discussed using matrices to systematize solving systems of coupled linear equations. Now we'll describe how to use Python to do this numerically. First a caveat: While generally much easier to do this numerically than "by hand", the numerical approach will not work for under-determined systems – unlike what we described earlier. (Note, Python installations include a "symbolic mathematics" package known as **sympy**. Using this package one *can* solve under-determined systems. However, a discussion of **sympy** is beyond the scope of this text.)

Let's use Python to solve the Eqs. 3.24.

```
1  ' ' '
2  Solving a system of linear coupled equations
3  ' ' '
4  import numpy as np
5  a = np.array([[2,3,1],[0,1,-2],[1,-4,-1]])
6  d = np.array([[11],[-4],[-10]]) # note: this is a column matrix
7  x = np.linalg.solve(a,d) # solve is in the linalg package in numpy
8  print("The solution vector is\n",x)
```

Listing 3.8: Sample code showing how to solve systems of coupled linear equations.

When I execute this code, I get

The solution vector is
[[1.]
[2.]
[3.]]

The solutions are the same as we obtained earlier using "manual" manipulation of the matrices, but with far less effort. Furthermore, the difficulty in solving coupled equations goes up as the square of the number of equations. But numerically, your only real limitation is computer time – which will certainly be much quicker than "human time"!

3.8.6 Using Python to Numerically Compute the Determinant of an Array

```
1  '''
2  Computing determinants
3  '''
4  import numpy as np
5  a = np.array([[7,1,9],[-2, 9, 0],[3, 1, 4]])
6  a_det = np.linalg.det(a) # need to use the linalg package in numpy
7
8  print(a, "\n")
9  print(a_det)
```
Listing 3.9: Sample code showing how to compute the determinant of an array.

When I execute this code, I get
[[7 1 9]
[-2 9 0]
[3 1 4]]

determinant = -0.9999999999999831
You can verify that, within round-off error, this is, indeed the determinant of the specified array.

3.8.7 Using Python to Compute the Eigenvalues and Eigenvectors of an Array

Recall that earlier in this chapter, we discussed deriving the eigenvalues, and then the eigenvectors of a matrix. For relatively small matrices, this is not terribly challenging. But the number of steps in this computation go as the square of the dimensionality of the space represented by the matrix. Using routines that come bundled with Python, the computation becomes absolutely trivial.

```
1  '''
2  Computing eigenvalues and eigenvectors
3  '''
4  import numpy as np
5  a = np.array([[1,2,3],[3,2,1],[1,0,-1]])
6  w, v = np.linalg.eig(a)   # eig is in the linalg sub-package of
7      numpy
8  print(np.round(w,3),'\n') # the "vector" array of eigenvalues
9
10 # the array of eigenvectors, one column for each eigenvalue:
11 print(np.round(v,3))
12
13 for i in range(3):
14     print("\nFor the", i, "eigenvalue =", np.round(w[i],3), \
15         ", the corresponding eigenvector is\n", \
16             np.round(v[:,i],3), "\n")
```
Listing 3.10: Sample code showing how to compute the eigenvalues and eigenvectors of an array.

When I execute this code, I get
[4.317 -2.317 0.]

[[0.584 0.736 0.408]
 [0.804 -0.382 -0.816]
 [0.11 -0.559 0.408]]

For the 0 eigenvalue = 4.317 , the corresponding eigenvector is
[0.584 0.804 0.11]

For the 1 eigenvalue = -2.317 , the corresponding eigenvector is
[0.736 -0.382 -0.559]

For the 2 eigenvalue = 0.0 , the corresponding eigenvector is
[0.408 -0.816 0.408]

You will note that the **eig** method returns *normalized* eigenvectors. That is the sum of the squares of the components of each eigenvector is 1. This particular piece of code looks more complicated than the other snippets in this chapter. In fact, the code is simple. What requires a bit of study is simply what we did to pick apart the arrays corresponding to the eigenvectors and the eigenvectors. Fortunately, the same tricks we used to pick apart lists also work with arrays. This concludes our discussion of matrices/arrays.

CHAPTER 4

Fourier Series

4.1 INTRODUCTION TO FOURIER SERIES

Fourier transforms are incredibly useful, but it will be difficult to convince you of this from just a single example. So bear with me as we examine several aspects of the "FT". We will start, not with the Fourier transform itself, but with the closely related Fourier series. Later, when we begin to cover the Fourier transform we will establish the connection between the two.

Suppose you have an infinite wave train, for example a square wave. You can always represent infinite, repeating wave trains as an infinite series of sines and/or cosines, having frequencies that are harmonics of some fundamental.

Amazingly, we can represent *any* repeating function, $g(x)$, by an infinite number of sines and cosines:

$$g(x) = \sum_{n=0}^{\infty} a_n \cos(n2\pi f_0 x) + \sum_{n=0}^{\infty} b_n \sin(n2\pi f_0 x) \ , \qquad (4.1)$$

where $f_0 = 1/T$ is the frequency, or one over the period, T, of the function $g(x)$.

The only problem now is to figure out the values of a_n and b_n. We start by multiplying both sides of Eq. 4.1 by $\cos(m2\pi f_0 x)$, where m is an integer, and then integrating over one full period, T. This gives us

$$\int_{-T/2}^{+T/2} g(x) \cos(m2\pi f_0 x)\, dx = \sum_{n=0}^{\infty} a_n \int_{-T/2}^{+T/2} \cos(n\, 2\pi f_0 x) \cos(m\, 2\pi f_0 x)\, dx$$
$$+ \sum_{n=0}^{\infty} b_n \int_{-T/2}^{+T/2} \sin(n\, 2\pi f_0 x) \cos(m\, 2\pi f_0 x)\, dx \ ,$$

where we have swapped the positions of the integral and summation symbols.

Now we need to think in terms of *even* and *odd* functions. If $g(x) = g(-x)$, then the function is said to be *even*. If $g(x) = -g(-x)$, then the function is said to be *odd*. Furthermore, the product of an even function with another even function gives us an even function and the product of an odd function with an odd function also gives us an even function. But the product of an odd function with an even function gives us an odd function. Clearly a cosine is even and a sine is odd. Therefore, the integrand

57

of the first integral on the right is an even function, while the integrand of the second integral on the right is an odd function. Here is a key point: *the integral of an odd function over an even interval (an interval symmetric about zero) is identically zero.* Therefore every term in the rightmost sum is zero, leaving us with:

$$\int_{-T/2}^{+T/2} g(x) \cos(m 2\pi f_0 x)\, dx = \sum_{n=0}^{\infty} \int_{-T/2}^{+T/2} a_n \cos(n\, 2\pi f_0 x) \cos(m\, 2\pi f_0 x)\, dx \ . \quad (4.2)$$

We'll see in a moment how to solve this expression for all the a_n, but first note that if we had instead multiplied Eq. 4.1 by $\sin(m\, 2\pi f_0 x)$ and integrated over one full period, all of the terms containing a_n would have vanished, leaving only the terms containing b_n:

$$\int_{-T/2}^{+T/2} g(x) \sin(m 2\pi f_0 x)\, dx = \sum_{n=0}^{\infty} \int_{-T/2}^{+T/2} b_n \sin(n\, 2\pi f_0 x) \sin(m\, 2\pi f_0 x)\, dx \ . \quad (4.3)$$

Let's solve Eq. 4.2 for a_n. First note that because $\cos(a) \cos(b) = \frac{1}{2}\left[\cos(a+b) + \cos(a-b)\right]$, Eq. 4.2 becomes

$$\int_{-T/2}^{+T/2} g(x) \cos(m 2\pi f_0 x)\, dx =$$

$$\frac{1}{2} \sum_{n=0}^{\infty} a_n \int_{-T/2}^{+T/2} \{\cos[2\pi f_0 x(n+m)] + \cos[2\pi f_0 x(n-m)]\}\, dx \ . \quad (4.4)$$

Now, the sums and differences of two integers will always be integers. Therefore, the integration on the right side will always be over an integral number of full periods – which is zero. (The integral of an integer number of periods of a cosine is always zero.) The only exception we have is when $n = m$ because then the argument of the second cosine is 1. In that case, the integral is just $a_m T/2$. So this means that

$$\int_{-T/2}^{+T/2} g(x) \cos(n 2\pi f_0 x)\, dx = a_n \frac{T}{2} \ ,$$

and therefore

$$a_n = \frac{2}{T} \int_{-T/2}^{+T/2} g(x) \cos(n 2\pi f_0 x)\, dx \ . \quad (4.5)$$

Returning to Eq. 4.4, we see that the special case of $m = n = 0$ leads to a different result because this puts zeros in the argument of all the cosines. Therefore in this case we get

$$a_0 = \frac{1}{T} \int_{-T/2}^{+T/2} g(x)\, dx = \langle g \rangle \ , \quad (4.6)$$

where $\langle g \rangle$ means the average value of g.

Pulling the same trick of multiplying Eq. 4.3 through with a sine and integrating over a period gives us

$$b_n = \frac{2}{T} \int_{-T/2}^{+T/2} g(x) \sin(n 2\pi f_0 x)\, dx \ . \quad (4.7)$$

Note that we do not have the same problem with b_0 as we had with a_0 because sine is an odd function and therefore taking the average value about the origin gives zero.

4.1.1 Example 1: The Square Wave

Now let's try this out on an actual function. Look at the square wave function plotted below by a simple Python program. Note the use of a "top hat" function to create the square wave. That is,

$$g(x) = \begin{cases} +h, & |x| \le \frac{T}{4}. \\ -h, & \frac{T}{2} \ge |x| < \frac{T}{4}. \end{cases} \tag{4.8}$$

Here is the code used to represent the function in Eq. 4.8:

```
'''
Square Wave Function: create it from a bunch of "top hat"
    functions
connected together.
'''
import numpy as np
import matplotlib.pyplot as plt
# need this for matplotlib to work in Jupyter:
%matplotlib inline

def top_hat(amplitude, width, x):
    if (np.abs(x) < width):
        return amplitude
    else:
        return -amplitude

a = 200.0 # width of pulse
h = 10.0  # amplitude of pulse
N = 1000  # total number of sample points
T = 2*a # the period
square_wave = []
x = np.linspace(-N/2, N/2, N) # N points, start at -N/2, end
    at +N/2
for i in x:
    square_wave.append(top_hat(h, a, (i+a/2)%T))
# connect a bunch of top hats. Note the use of %
# And now plot it
plt.figure(1)
plt.plot(x,square_wave, color = "blue")
plt.xlabel('x (arb)')
plt.ylabel('Amplitude (arb)')
plt.ylim(-1.1*h, 1.1*h)
plt.title('Square Wave')
plt.grid(True)
plt.show()
```

Listing 4.1: Code for plotting the square wave from Eq. 4.8.

And here is the resulting plot:

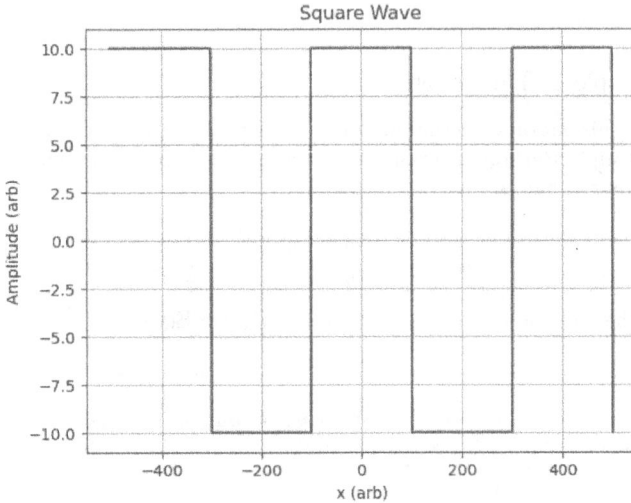

FIGURE 4.1: A plot of the square wave of Eq. 4.8.

This is clearly an even function: it is completely symmetrical about the origin. We can therefore use Eqs. 4.5 and 4.6 to deduce a_n and not even worry about b_n. Let's use Eq. 4.6 first. In this case, the average value of $g(x)$ is zero, so $a_0 = 0$.

Now plugging the square wave function into Eq. 4.5 gives us

$$a_{n \neq 0} =$$

$$\frac{2}{T} \int_{-T/2}^{-T/4} -h \cos\left(\frac{n2\pi x}{T}\right) dx + \frac{2}{T} \int_{-T/4}^{+T/4} h \cos\left(\frac{n2\pi x}{T}\right) dx +$$

$$\frac{2}{T} \int_{+T/4}^{+T/2} -h \cos\left(\frac{n2\pi x}{T}\right) dx .$$

Or, making use of symmetry,

$$a_{n \neq 0} = \frac{4h}{T} \left[\int_{0}^{+T/4} \cos\left(\frac{n2\pi x}{T}\right) dx - \int_{T/4}^{T/2} \cos\left(\frac{n2\pi x}{T}\right) dx \right] .$$

Then,

$$a_{n \neq 0} = \frac{4h}{n\pi} \sin\left(\frac{n\pi}{2}\right) .$$

Let's plot Eq. 4.1 with these values of a_n and for $b_n = 0$. Clearly I can't include an infinite number of terms, as indicated in Eq. 4.1, but let's see what happens as I increase the number of terms, from 5 to 9 to 50. First the Python code:

```
1  '''
2  Fourier Series Approximation of the Square Wave Function
3  '''
4  import numpy as np
5  import matplotlib.pyplot as plt
6
7  # need this for matplotlib to work in Jupyter:
8  %matplotlib inline
9
10 # a square wave is made up of "top hat" fcts:
11 def top_hat(amplitude, width, x):
12     if (np.abs(x) < width):
13         return amplitude
14     else:
15         return -amplitude
16
17 def A(n, h):  # the expansion coefficients
18     if (n == 0):
19         return 0.0  # from Eq. 7
20     else:
21         return 4.0*h/(n*np.pi)*np.sin(n*np.pi/2.0)  # from Eq. 12
22
23 w = 200.0 # width of pulse
24 h = 10.0  # amplitude of pulse
25 N = 1000  # total number of sample points
26 T = 2*w   # the period
27 square_wave = []
28 five_term = []
29 nine_term = []
30 fifty_term = []
31 a = []
32
33 for j in np.linspace(0,50,51):
34     a.append(A(j,h))
35
36 # start at -N/2, end at +N/2, inc by 1
37 x = np.linspace(-float(N)/2, float(N)/2, N)
38
39 # here's how to make a square wave from top hats
40 for i in x:
41     square_wave.append(top_hat(h, w, (i+w/2)%T))
42     tmp = 0
43     for m in range(0,5):
44         tmp += a[m]*np.cos(2*np.pi*m*i/T)
45     five_term.append(tmp)
46
47     tmp = 0
48     for m in range(0,9):
49         tmp += a[m]*np.cos(2*np.pi*m*i/T)
50     nine_term.append(tmp)
51
52     tmp = 0
53     for m in range(0,50):
54         tmp += a[m]*np.cos(2*np.pi*m*i/T)
55     fifty_term.append(tmp)
56
```

```
57 # And now plot it
58 plt.figure()
59 plt.plot(x,square_wave, label = "Square Wave", color = "black")
60 plt.plot(x,five_term, label = "5 Terms", color = "red")
61 plt.plot(x,nine_term, label = "9 Terms", color = "green")
62 plt.plot(x,fifty_term, label = "50 Terms", color = "blue")
63 plt.xlabel('x (arb)')
64 plt.ylabel('Amplitude (arb)')
65 plt.ylim(-1.4*h, 1.4*h)
66 plt.title('Square Wave')
67 plt.legend(bbox_to_anchor=(1.05, 1), loc=0, borderaxespad=0.)
68 plt.grid(True)
69 plt.show()
```

Listing 4.2: Code for plotting the square wave from Eq. 4.8, along with various approximations from a Fourier series.

And now the plot:

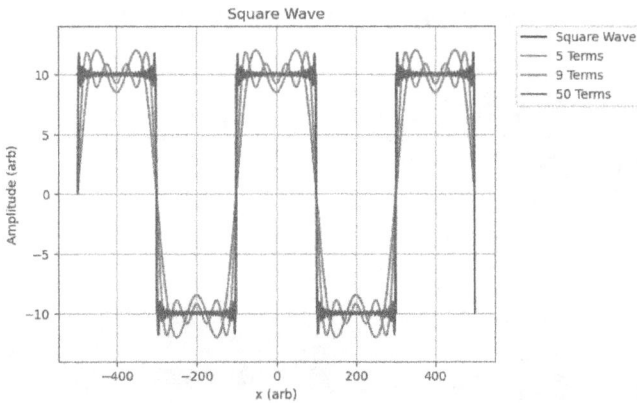

FIGURE 4.2: A plot of the square wave of Eq. 4.8 and approximations to a square wave using Fourier series.

4.1.2 Example 2: The Sawtooth Wave

Consider the sawtooth wave. To be specific, let's define a single "tooth" of a sawtooth wave as

$$g(x) = \left\{ \tfrac{2h}{T}x \ , \quad -\tfrac{T}{2} \le x \le \tfrac{T}{2} \right. . \tag{4.9}$$

Then, between $-T/2$ and $+T/2$, the wave is a straight line having a slope of $2h/T$. It should also be centered horizontally such that the midpoint of the sloped line is at $x = 0$. Let's plot this function:

```
1 '''
2 Plot of the Sawtooth Function
3 '''
4 import numpy as np
5 import matplotlib.pyplot as plt
```

```
6
7  # need this for matplotlib to work in Jupyter:
8  %matplotlib inline
9
10 # a sawtooth wave is made up of many "teeth":
11 def tooth(amplitude, width, x):
12     return (1.0*amplitude)/(1.0*width)*x
13
14 w = 200.0    # width of pulse
15 h = 10.0     # height of pulse
16 N = 1000     # total number of sample points
17 T = 2*w      # the period
18 saw_tooth_wave = []
19
20 # N points, start at -N/2, end at +N/2:
21 x = np.linspace(-float(N)/2, float(N)/2, N)
22
23 # here's how to make a sawtooth from teeth
24 for i in x:
25     saw_tooth_wave.append(tooth(h, w, (i+w)%T)-h)
26
27 # And now plot it
28 plt.figure()
29 plt.plot(x,saw_tooth_wave, label = "Sawtooth Wave",
30     color = "blue")
31 plt.xlabel('x (arb)')
32 plt.ylabel('Amplitude (arb)')
33 plt.ylim(-1.2*h, 1.2*h)
34 plt.title('Sawtooth Wave')
35 plt.grid(True)
36 plt.show()
```

Listing 4.3: Code for plotting the sawtooth wave from Eq. 4.9.

Note that the function is *odd*, which means that we only need to find b_n. Then, from Eq. 4.7,

$$b_n = \frac{2}{T} \int_{-T/2}^{+T/2} g(x) \sin\left(\frac{n2\pi}{T}x\right) dx =$$

$$\frac{2}{T}\frac{2h}{T} \left\{ \int_{-T/2}^{0} -x \sin\left(\frac{n2\pi}{T}x\right) dx + \int_{0}^{T/2} x \sin\left(\frac{n2\pi}{T}x\right) dx \right\} .$$

For

$$a \equiv \frac{n2\pi x}{T} .$$

Integration by parts gives us

$$b_n = \frac{4h}{T^2} \left\{ -\left(\frac{\sin(ax)}{a^2} - \frac{x\cos(ax)}{a}\right)\Big|_{-T/2}^{0} + \left(\frac{\sin(ax)}{a^2} - \frac{x\cos(ax)}{a}\right)\Big|_{0}^{T/2} \right\} .$$

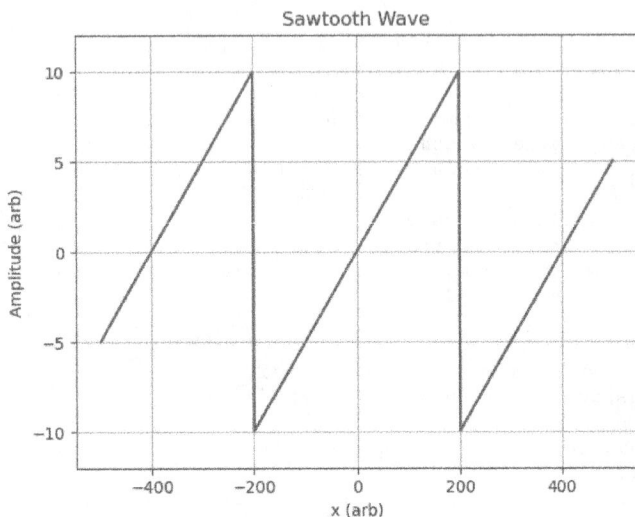

FIGURE 4.3: A plot of the sawtooth wave of Eq. 4.9.

The sine terms vanish, as do the cosine terms when the limit $= 0$, leaving us with

$$b_{n\neq0} = -\frac{2h}{n\pi}(-1)^n \ ,$$

and

$$b_{n=0} = 0 \ .$$

We can easily generate these coefficients and plot the Fourier series approximation to the sawtooth:

```
1  '''
2  Fourier Series Approximation of the Sawtooth Function
3  '''
4  import numpy as np
5  import matplotlib.pyplot as plt
6
7  # need this for matplotlib to work in Jupyter:
8  %matplotlib inline
9
10 # a sawtooth wave is made up of many "teeth"
11 def tooth(amplitude, width, x):
12     return (1.0*amplitude)/(1.0*width)*x
13
14 def B(n, h):  # the expansion coefficients
15     if (n == 0):
16         return 0.0
17     else:
```

```
18          return −2*h/(n*np.pi)*(−1)**n # from Eq. 17
19
20 w = 200.0    # width of pulse
21 h = 10.0     # height of pulse
22 N = 1000     # total number of sample points
23 T = 2*w      # the period
24 saw_tooth_wave = []
25 one_term = []
26 three_term = []
27 fifty_term = []
28 b = []
29
30 for j in np.linspace(0,50,51):
31     b.append(B(j,h))
32
33 # N points, start at −N/2, end at +N/2:
34 x = np.linspace(−float(N)/2, float(N)/2, N)
35
36 # here's how to make a sawtooth from a tooth
37 for i in x:
38     saw_tooth_wave.append(tooth(h, w, (i+w)%T)−h)
39
40     tmp = 0.0
41     for m in range(1,2):
42         tmp += b[m]*np.sin(2*np.pi*m*i/T)
43     one_term.append(tmp)
44
45     tmp = 0.0
46     for m in range(1,4):
47         tmp += b[m]*np.sin(2*np.pi*m*i/T)
48     three_term.append(tmp)
49
50     tmp = 0.0
51     for m in range(1,50):
52         tmp += b[m]*np.sin(2*np.pi*m*i/T)
53     fifty_term.append(tmp)
54
55 # And now plot it
56 plt.figure()
57 plt.plot(x,saw_tooth_wave, label = "Sawtooth Wave",
58     color = 'black')
59 plt.plot(x,one_term, label = "1 Term", color = "red")
60 plt.plot(x,three_term, label = "3 Terms", color = "green")
61 plt.plot(x,fifty_term, label = "50 Terms", color = "blue")
62 plt.xlabel('x (arb)')
63 plt.ylabel('Amplitude (arb)')
64 plt.ylim(−1.2*h, 1.2*h)
65 plt.title('Sawtooth Wave')
66 plt.legend(bbox_to_anchor=(1.05, 1), loc=0, borderaxespad=0.)
67 plt.grid(True)
68 plt.show()
```

Listing 4.4: Code for plotting the sawtooth wave from Eq. 4.9, as well as approximations from the Fourier Series.

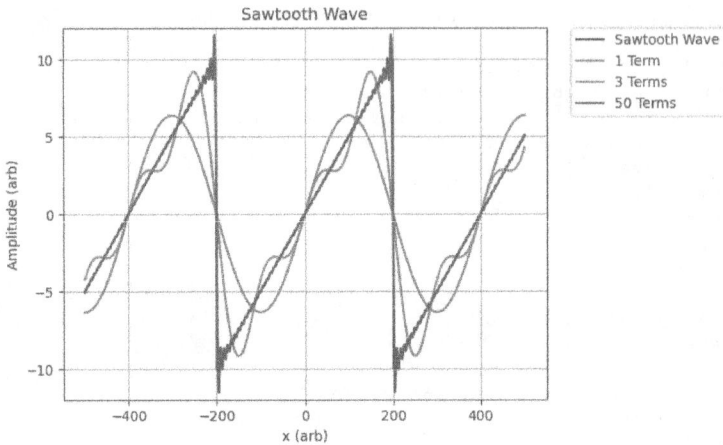

FIGURE 4.4: A plot of the sawtooth wave of Eq. 4.9 and approximations to a sawtooth using Fourier series.

Homework 4.1: The Triangle Wave

See if you can derive the coefficients for a triangle wave. Hint: Place the peak of a triangle on $x = 0$; then your function will be even and will only require a_n coefficients. Then write a program that plots the triangle wave and successive approximations of it using the Fourier series.

4.2 SUMMARY

All of our examples were of waveforms that were either odd or even, but this needn't be the case. For example, if you were to add a sawtooth wave to a square wave, the result must contain both a_n and b_n coefficients and therefore cannot be strictly even or odd. Not a problem: just use Eqs. 4.5 and 4.6 to solve for a_n and Eq. 4.7 to solve for b_n.

You may be tempted to conclude that the purpose of Fourier series analysis is in order to approximate periodic functions. This is partially correct: suppose you wish to create a sawtooth wave from rf sources. You can do a pretty good job using just one rf frequency added – in the correct ratios given by b_n – to a few harmonics of that fundamental frequency. Think about that: you can create any repetitive waveform you wish from the appropriately weighted fundamental frequency plus a handful of its harmonics. That's powerful stuff!

But another reason for this sort of analysis is to see what are the important frequency components in any repetitive signal, be it acoustic, electric/electromagnetic, or optical. This sort of knowledge can help you by telling you what frequencies you may need to filter out or enhance, depending on your ultimate purpose.

Finally, we have made a big deal about the fact that these waveforms must be repetitive. But by tweaking Eq. 4.1, specifically by converting sums into integrals, we can make a much more general tool – Fourier transforms – that will allow us to seamlessly move between *conjugate variable pairs* such as time and frequency or space and momentum. Fourier transforms are the subject of the next chapter.

## CHAPTER 5

Fourier Transforms

5.1 GENERALIZING THE FOURIER SERIES

Suppose you have an infinite periodic wave train, for example a square wave. As we have seen, you can always represent this waveform as an infinite series of sines and/or cosines, having frequencies that are harmonics of some fundamental. But what if your function is not periodic? What if it does not repeat? You can still deduce the frequency content of that function. In this chapter we will generalize the Fourier series analysis to such non-periodic functions.

5.1.1 Integral Transforms

Suppose you have data that consist of a series of values as a function of time but that, unlike an "infinite square wave", do not repeat. For example, the noise "spectrum" in a supposedly quiet room, or an EEG (Electroencephalography, or "brain waves") signal as a function of time. Or how about the temperature of a certain place in the Pacific Ocean as a function of time? Perhaps you care about the seismic vibrations following an earthquake. Even stock market fluctuations versus time could be interesting data. In all these examples an important aspect of the measured signal is how it varies with time. Indeed, the independent variable need not be time: consider the variation of the number density of galaxies as a function of the angular spread across the sky; is there some sort of regularity in that pattern?

The mathematical process of determining the "frequency components" of a signal as a function of that signal's independent variable, whether it be time, space, or any other variable, is referred to as the *Fourier Transform*. In the following equations, we show this specific mathematical procedure.

Recall that a periodic function can be expanded in terms of sines and cosines of harmonics of a fundamental frequency:

$$F(t) = \sum_{n=0}^{\infty} a_n \cos(n 2\pi f_0 t) + \sum_{n=0}^{\infty} b_n \sin(n 2\pi f_0 t).$$

If our function is not periodic, we can follow the same idea, but we have to go to the limit of the period going to infinity and we have to include all frequencies, not just integer multiples of some fundamental. Thus the sums become integrals and the limits go to $\pm\infty$:

$$F(t) = \int_{-\infty}^{\infty} a(\nu) \cos(2\pi\nu t)\, d\nu + \int_{-\infty}^{\infty} b(\nu) \sin(2\pi\nu t)\, d\nu.$$

This is equivalent to

$$F(t) = \int_{-\infty}^{\infty} r(\nu) \cos(2\pi\nu t + \phi(\nu))\, d\nu \quad , \quad \text{or}$$

or

$$F(t) = \int_{-\infty}^{\infty} \Phi(\nu) e^{-2\pi i \nu t}\, d\nu. \tag{5.1}$$

Equation 5.1 can be multiplied on both sides by $e^{2\pi i \nu t'}$ and integrated over t to give

$$\Phi(\nu) = \int_{-\infty}^{+\infty} F(t) e^{2\pi i \nu t}\, dt. \tag{5.2}$$

Equation 5.1 takes us from, say, the frequency domain into the time domain, and Eq. 5.2 takes us from the time domain back into the frequency domain.

$\Phi(\nu)$ and $F(t)$ are said to be *Fourier pairs* and we can symbolically show their connection by writing

$$\Phi(p) \rightleftharpoons F(x)$$

where p and x can be any pair of *conjugate variables*, including frequency and time, momentum and length, *etc.*

Let's consider a specific example. Suppose you're looking at an EEG signal. Scientists have determined that different frequency ranges correspond to different cognitive activities. For example, "theta waves" (frequency range 4-7 Hz) are associated with inhibition of elicited responses. That is, the electromagnetic activity in the 4-7 Hz range spikes when a person is actively trying to repress a responsive action. On the other hand, "beta waves" (16-31 Hz) are associated with active thinking, focus, high alertness, and anxiety. In order to parse the EEG signals into the various frequency bands, you can take a Fourier transform.

To make this specific example simpler to conceptualize, let's manufacture some data, using the following function:

$$F = \sin\left(2\pi f_1 x\right) + \frac{1}{2}\sin\left(2\pi f_2 x\right) \quad , \tag{5.3}$$

where f_1 and f_2 are some arbitrarily chosen frequencies. In the following code, I chose $f_1 = 50$ Hz and $f_2 = 80$ Hz. Then, using Eq. 5.3, we can manufacture some "data". A plot of those data is shown in the upper graph of Fig. 5.1.

Then, using the function `fft`, we take the Fourier transform of those data in order to determine their frequency content. The Fourier transform of those data is shown in the lower graph of Fig. 5.1.

Details of the code follow:

In lines 4-7 we load in the packages we will need. The Fourier transform and inverse transform packages, corresponding to Eqs. 5.1 and 5.2, respectively, are named

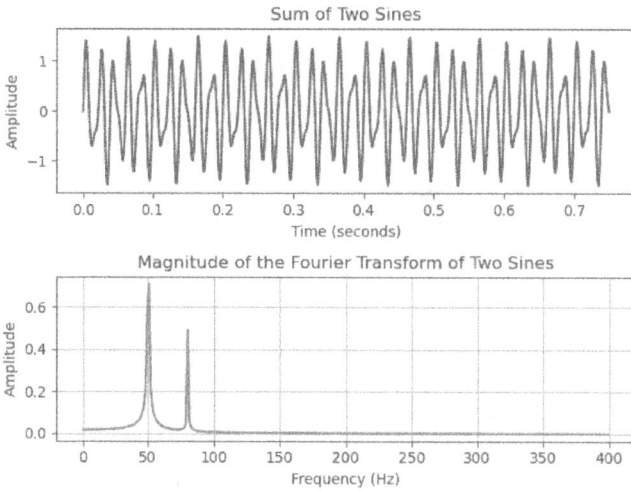

FIGURE 5.1: Upper graph: A plot of the sum of two sines. Lower graph: a plot of the Fourier transform of the function in the upper figure.

fft and ifft and are part of the scipy package. We also load numpy and matplotlib, as usual.

In lines 17-20 we define x and $F(x)$. (I used different names for the variables in the code, as compared to those in Eq. 5.1 to emphasize that you can name your Python variables however you like.) Notice that in line 20 $F(x)$ is defined to be the sum of two sines having different frequencies and amplitudes. Specifically, a sinusoid of amplitude $= 1$ and having a frequency of 50 (angular frequency $= 2\pi \times 50$) is added to a sinusoid of amplitude $1/2$ and having a frequency of 80. If x is a time and has units of seconds, then the frequencies will have units of Hz, as I have already indicated above.

We then take the Fourier transform of F using the fft function in line 21. It's just that simple. Notice that fft returns an array and that array has the same length as the array that was fft's argument (F).

Before we continue with the code, jump ahead to look at the second graph in the output, that is, the Fourier transform of the "data". You see peaks at frequencies of 50 and 80 – as we should expect, since the "data" were made from the sum of two sine functions at those frequencies.

In line 23 we define the independent variable in the transform space, putting it into the array named p. Notice the inverse relationship between Δx in the "time domain" compared with Δp in the "frequency domain". You may also have noticed that there are only $N/2$ values in p but N values in x. Basically, we have simply not shown the entire Fourier transform. Try changing the code by removing the "$//2$" in line 23 and the corresponding "$/2$" in line 38. (To keep the "frequency" scale correct, you'll also have to change the "2.0*" in line 23 to "1.0*".) Now run the code again and see what you have. You should see the two peaks we saw before, again at $p = 50$ and $p = 80$, but we also see similar looking peaks at $(800-50)$ and $(800-80)$ as well.

These are so-called "negative frequency" results and are usually ignored. They don't really have a physical meaning, but are the mathematical result of using a complex exponential in Eq. 5.1, rather than a sine and/or cosine. If you were to analytically integrate Eq. 5.1 for $F(x) = \sin(2\pi x t)$ you'd get a *pair* of delta functions, one at $+2\pi x$ and one at $-2\pi x$. If the math is beyond your current understanding, don't worry about it: your math knowledge will grow. In the meantime, just do what most do: only plot the positive half of the Fourier transform spectrum.

Notice, by the way, that in line 38 we plotted the *magnitude* of Phi, using the **abs** function from **numpy**. In general, the Fourier transform of a function gives an array of complex numbers. The physical significance of this is that a signal can be expressed as a sum of sinusoids of different frequencies *and phases*. If we were to express each complex value in the Fourier transform array as a magnitude and a phase, you'd see the phase associated with each frequency. (In this example, we can see by how we constructed the "signal data" that the phases are all zero, and hence uninteresting to plot.)

```
1  '''
2  Example of numerical Fourier Transform
3  '''
4  import numpy as np
5  from scipy.fftpack import fft
6  from scipy.fftpack import ifft   # Inverse FFT, not used here
7  import matplotlib.pyplot as plt
8
9  # need this for matplotlib to work in Jupyter:
10 %matplotlib inline
11
12 # Number of sample points
13 N = 600
14 # sample spacing
15 T = 1.0 / 800.0 # let's say T is in seconds
16 # N points, start at 0, end at N*T. x is in seconds
17 x = np.linspace(0.0, N*T, N)
18 freq1 = 50.0    # freq is in 1/sec or Hz
19 freq2 = 80.0
20 F = np.sin(freq1 * 2.0*np.pi*x) + 0.5*np.sin(freq2 * 2.0*np.pi*x)
21 Phi = fft(F)
22 # "//" forces integer division in Python 3.x
23 p = np.linspace(0.0, 1.0/(2.0*T), N//2)
24
25 # And now plot the Fourier transform
26 plt.figure()
27 # make 2 plots vertically, and 1 horizontally.
28 # All plot commands are for the first plot until
29 # we give the next subplot command:
30 plt.subplot(211)
31 # plot the time-domain data:
32 plt.plot(x,F, label = "Time Domain", color = "blue")
33 plt.xlabel('Time (seconds)')
34 plt.ylabel('Amplitude')
35 plt.title('Sum of Two Sines')
36
37 plt.subplot(212)
```

```
38  plt.plot(p, 2.0/N * np.abs(Phi[0:int(N/2)]), \
39          label = "Frequency Domain", color = "red")
40  plt.xlabel('Frequency (Hz)')
41  plt.ylabel('Amplitude')
42  plt.title('Magnitude of the Fourier Transform of Two Sines')
43  plt.grid(True)
44  #plt.legend(bbox_to_anchor=(1.05, 1), loc=0, borderaxespad=0.)
45  plt.tight_layout()
46  plt.show()
```

Listing 5.1: Code for computing and plotting the Fourier transform of a manufactured function.

5.1.2 The "Fast" Fourier Transform

You may have wondered why the Fourier transform function in the above code was named `fft` rather than just `ft`. The extra `f` stands for "fast". It turns out that if you have precisely 2^n points in the data you want to transform, there is an algorithm that can do the transformation far more efficiently than the brute force numerical integration of Eq. 5.1. Nearly everyone who makes use of Fourier transforms uses this algorithm – including us. But wait: We only had 600 points in our "signal" data! For the `fft` algorithm to be applicable we had to have either 512 or 1024 points (2^9 or 2^{10}). What gives? What gives is that the good people who wrote scipy recognized the inconvenience of having precisely 2^n points and automatically pad your data with zeros on either side of the numbers in the array you use as the argument in `fft`. Nice.

Another restriction inherent in the numerical FFT (our new abbreviation for the Fast Fourier Transform) is that input data must be equally spaced. If your data, perhaps taken from some experiment are *not* equally spaced, you will have to either fit your data to a function or interpolate your data in order to place the input to `fft` on a regular grid. **Numpy** has built-in functions that do this for you.

5.1.3 Other Restrictions on Taking Fourier Transforms

There are other restrictions on taking a Fourier transform, whether of the "fast" type or not. These are referred to a the *Dirichlet conditions*. They are:

- The function must be square-integrable. That is, the function $|F(x)|^2$ must exist and be finite.
- $F(x)$ and $\Phi(p)$ must be single-valued.
- $F(x)$ and $\Phi(p)$ must be *piece-wise continuous*. The function can be broken up into pieces, but the pieces cannot contain singularities.
- $F(x)$ and $\Phi(p)$ have upper and lower bounds. Technically this is a sufficient but not necessary condition: as a simple example, the delta function violates this condition.

5.2   FOURIER TRANSFORMS OF KEY FUNCTIONS

5.2.1   The Top-Hat Function

The top-hat function is given by

$$\Pi_a(x) = \begin{cases} 0 \ , & -\infty < x < -a/2 \\ 1 \ , & -a/2 < x < +a/2 \\ 0 \ , & +a/2 < x < \infty \ . \end{cases} \tag{5.4}$$

We can easily compute the FT of the Top Hat function analytically. You should end up with

$$\Phi(p) = a \left\{ \frac{\sin(\pi pa)}{\pi pa} \right\} \equiv a\,\mathrm{sinc}(\pi pa).$$

In the program below we create a Top Hat function and take and plot its FT. Notice that the smaller a is, the broader the *sinc* function. You can show that the *sinc* has zero crossings at $p = \pm 1/a, \pm 2/a, \pm 3/a, ...$ By the way, notice that I played fast and loose when I plotted the result. Try plotting the FT in the "conventional" way to see what you get. Convince yourself that the result is consistent with the theoretical one.

5.2.2   The Sinc Function

If the Fourier transform of the Top Hat function is a *sinc*, what do you think the FT of a *sinc* is? The analytic derivation can be done, but it is rather painful. Rather than write a whole new piece of code to verify your guess through derivation or writing a whole new program, use the symmetry of the FT versus FT^{-1}. That is, just take the inverse FT of the sync you came up with below.

```
1  '''
2  Top Hat and sinc Demo
3  '''
4  import numpy as np
5  from scipy.fftpack import fft
6  from scipy.fftpack import ifft # in case I need it!
7  import matplotlib.pyplot as plt
8
9  # need this for matplotlib to work in Jupyter:
10 %matplotlib inline
11
12 def topHat(a, x):
13     if (x < -a/2):
14         return 0.0
15     if (x > a/2):
16         return 0.0
17     return 1.0
18
19 a = 8.0
20 N = 512 # number of sample points
21
22 T = 1.0 / 4.0 # sample spacing
```

```
23  # N points, start at 0, end at N*T. x is in seconds:
24  x = np.linspace(-int(N*T/2), int(N*T/2), N)
25  TH = []
26  for i in x:
27      TH.append(topHat(a, i))
28
29  Phi = fft(TH)
30  p = np.linspace(-1.0/(4.0*T), +1.0/(4.0*T), int(N/2))
31
32  # And now plot the Fourier Transform
33  plt.figure()
34  # make 2 plots vertically, and 1 horizontally.
35  # All plot commands are for the first plot until
36  # we give the next subplot command:
37  plt.subplot(211)
38  # plot the time-domain data:
39  plt.plot(x,TH, label = "Time Domain", color = "blue")
40  plt.xlabel('Time (seconds)')
41  plt.ylabel('Amplitude')
42  plt.ylim(-0.1, 1.1)
43  plt.title('Top Hat Function')
44
45  plt.subplot(212)
46  plt.plot(p[0:int(N/4)], 2.0/N * np.abs(Phi[int(3*N/4):N]), \
47          label = "Frequency Domain", color = "red")
48  plt.plot(p[int(N/4):int(N/2)], 2.0/N * np.abs(Phi[0:int(N/4)]), \
49          color = "red")
50  plt.xlabel('Frequency (Hz)')
51  plt.ylabel('Amplitude')
52  plt.title('Magnitude of the Fourier Transform of a Top Hat
53      Function')
54  plt.grid(True)
55  #plt.legend(bbox_to_anchor=(1.05, 1), loc=0, borderaxespad=0.)
56  plt.tight_layout()
57  plt.show()
```

Listing 5.2: Code for computing and plotting the Fourier transform of the Top Hat function of Eq. 5.4.

5.2.3 The Gaussian Function

The gaussian is one of the most useful non-trigonometric functions in physics. Suppose we express it as

$$G(A, x, \sigma) \equiv A \exp\left(-\frac{x^2}{\sigma^2}\right) , \tag{5.5}$$

where σ is the "width parameter". Note that if we want the gaussian to be normalized, that is if we wish the integral (from $-\infty$ to $+\infty$) of the gaussian to be 1, then $A = (\sigma\sqrt{\pi})^{-1}$.

You can "easily" take the analytic FT of a gaussian, normalized or not, by "completing the square". Doing so gives you

$$\tilde{g}(p) = A\sigma\sqrt{\pi}e^{-\pi^2\sigma^2 p^2} , \tag{5.6}$$

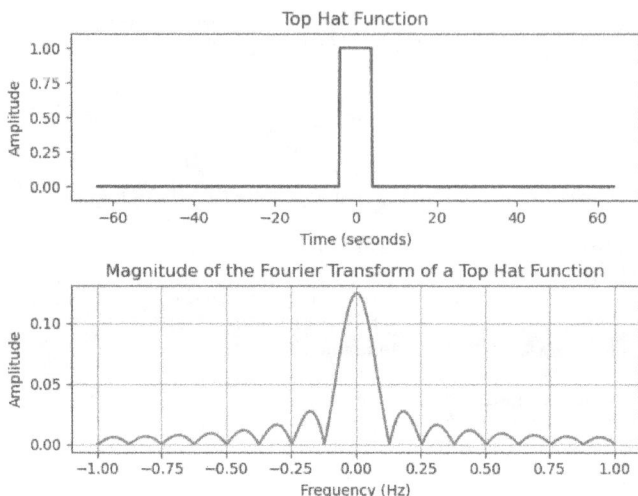

FIGURE 5.2: Upper graph: a plot of a Top Hat function. Lower graph: a plot of the Fourier transform of the Top Hat function in the upper figure.

where we have used the "tilde convention" to explicitly indicate a Fourier transform. Notice that if the gaussian had been normalized we would have had the particularly symmetric result

$$e^{-\pi^2 \sigma^2 p^2} = \text{FT}\left(\frac{1}{\sigma\sqrt{\pi}} e^{-\frac{x^2}{\sigma^2}}\right).$$

This is a very interesting result: The FT of a gaussian is another gaussian. Furthermore, the width of the FT gaussian is just $1/(\pi\sigma)$. That is, it is proportional to the inverse of the width of the original gaussian. Thus, the wider a gaussian, the more narrow its Fourier transform; the more narrow the gaussian, the wider its Fourier transform. Let's write some code to generate a gaussian and then take its Fourier transform. What happens to the Fourier transform if you have an "offset"? That is, what happens if you call $G(A, x - x_0, \sigma)$, where x_0 is a constant? (Try adding a plot of the *phase* of the Fourier transform to the graph that shows the magnitude.)

5.2.4 The Exponential Decay Function

Consider the decay function

$$f(x) = e^{-|x|/\sigma}.$$

This function is used to describe, among other things, the spontaneous decay of the excited state of nuclei, atoms, molecules, etc. The absolute value comes in because when considering decay, we generally only consider positive times. Then, the Fourier transform becomes

$$\tilde{f}(p) = \int_0^\infty e^{-x/\sigma} e^{i2\pi px} dx \ ,$$

where because we are only interested in $x \geq 0$, we can use 0 as our integral's lower bound and dispense with the absolute values. The integral is straight-forward, but leaves us with the complex result:

$$\tilde{f}(p) = \frac{\sigma}{1 + 4\pi^2 \sigma^2 p^2} + i\frac{2\pi\sigma^2 p}{1 + 4\pi^2 \sigma^2 p^2}.$$

or

$$\left|\tilde{f}(p)\right|^2 = \frac{\sigma^2}{1 + 4\pi^2 \sigma^2 p^2} \ , \quad \phi(p) = \arctan\left(\frac{1}{2\pi\sigma p}\right).$$

In the first equation we expressed $\tilde{f}(p)$ in terms of its real and imaginary parts and in the second set of equations, we expressed $\tilde{f}(p)$ in terms of its magnitude and phase. Both representations have their uses: if the exponential decay is used to describe, for example, the decay of an excited state of an atom, the real part (as well as the magnitude, which differ only by a multiplicative constant) are the equations of a *Lorentzian* and represent the frequency distribution of the spontaneously emitted light. Note that in the real part of the FT, the longer the "lifetime", the narrower the "linewidth"; the shorter the lifetime, the broader the frequency distribution. The imaginary part of the FT, when applied to decaying atoms, is related to the *index of refraction* of the system which, in turn, is related to the momentum of the photons being emitted. Lots of physics in this example!

Since the Fourier transform was so simple to do analytically, we dispense with the **fft** and just directly plot the above functions.

```
1  '''
2  Exponential Decay
3  '''
4  import numpy as np
5  import matplotlib.pyplot as plt
6
7  # need this for matplotlib to work in Jupyter:
8  %matplotlib inline
9
10 def decay(x, sigma):
11     return np.exp(-abs(x)/sigma)
12
13 def decReal(p, sigma):
14     denom = 1.0 + (2*np.pi*sigma*p)**2
15     return sigma/denom
16
17 def decImag(p, sigma):
18     denom = 1.0 + (2*np.pi*sigma*p)**2
19     return 2*np.pi*sigma*sigma*p/denom
20
21 def decMag(p, sigma):
22     denom = 1.0 + (2*np.pi*sigma*p)**2
```

```
23       return sigma*sigma/denom
24
25  sigma = 2.0
26  N = 500 # number of sample points
27
28  T = 1.0 / 8.0 # sample spacing
29  # N points, start at 0, end at N*T. x is in seconds:
30  x = np.linspace(0, N*T, N)
31  EDecay = []
32  fTildeReal = []
33  fTildeImag = []
34  fTildeMag = []
35  for i in x:
36      EDecay.append(decay(i, sigma))
37
38  p = np.linspace(-1.0/(4.0*T), +1.0/(4.0*T), N)
39  for j in p:
40      fTildeReal.append(decReal(j, sigma))
41      fTildeImag.append(decImag(j, sigma))
42      fTildeMag.append(decMag(j, sigma))
43
44  # And now plot the Fourier Transform
45  plt.figure()
46  # make 2 plots vertically, and 1 horizontally.
47  # All plot commands are for the first plot until
48  # we give the next subplot command:
49  plt.subplot(211)
50  # plot the time-domain data:
51  plt.plot(x[:int(N/4)],EDecay[:int(N/4)], \
52          label = "Time Domain", color = "blue")
53  plt.xlabel('Time (seconds)')
54  plt.ylabel('Amplitude')
55  plt.ylim(-0.1, 1.1)
56  plt.title('Exponential Decay Function')
57
58  plt.subplot(212)
59  plt.plot(p[int(N/4):int(3*N/4)], fTildeReal[int(N/4):int(3*N/4)],\
60          label = "Real", color = "red")
61  plt.plot(p[int(N/4):int(3*N/4)], fTildeImag[int(N/4):int(3*N/4)],\
62          label = "Imaginary", color = "green")
63  plt.plot(p[int(N/4):int(3*N/4)], fTildeMag[int(N/4):int(3*N/4)],\
64          label = "Magnitude", color = "blue")
65  plt.xlabel('Frequency (Hz)')
66  plt.ylabel('Amplitude')
67  plt.title('Fourier Transform of a Decay Function')
68  plt.ylim(-1.5, sigma*sigma*1.1)
69  plt.grid(True)
70  plt.legend(bbox_to_anchor=(1.05, 1), loc=0, borderaxespad=0.)
71  plt.tight_layout()
72  plt.show()
```

Listing 5.3: Code to plot a decaying exponential and its Fourier transform.

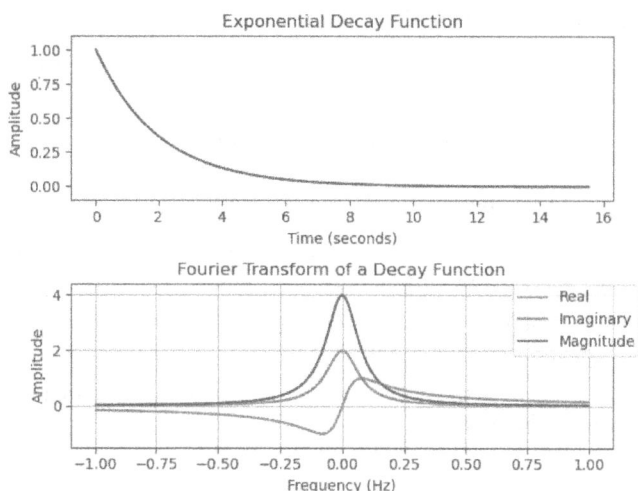

FIGURE 5.3: Upper graph: a plot of a decaying exponential. Lower graph: a plot of the Fourier transform of the decaying exponential in the upper figure.

5.2.5 The Dirac Delta Function

We've already mentioned the Dirac delta function. When combined with Fourier transforms its chief use is in deducing the FTs of other, more complicated functions. The Dirac delta has the following properties:

- $\delta(x) = 0$ unless $x = 0$. Otherwise $\delta(0) = \infty$.
- $\int_{-\infty}^{\infty} \delta(x)dx = 1$.
- $\delta(x) \rightleftharpoons 1$.
- $\delta(x - a) = 0$, unless $x = a$, $\int_{-\infty}^{\infty} f(x)\delta(x - a)dx = f(a)$.

From this last property,

$$\int_{-\infty}^{\infty} e^{i2\pi px}\delta(x - a)dx = e^{i2\pi pa}.$$

and from this, for the special case of $a = 0$, we get the second property above.

5.2.6 A Pair of Dirac Delta Functions

Suppose we have a pair of delta functions, situated symmetrically on either side of the origin. Then we see from the last result that the Fourier transform of this pair is a cosine:

$$\delta(x - a) + \delta(x + a) \rightleftharpoons e^{i2\pi pa} + e^{-i2\pi pa} = 2\cos(2\pi pa).$$

5.2.7 A "Comb" of Dirac Delta Functions

Finally, let's consider a "comb" of an infinite number of uniformly spaced δ-functions:

$$\text{III}_a(x) = \sum_{n=\infty}^{\infty} \delta(x - na).$$

This is a useful topic because

1. It allows us to make a connection between the general Fourier transform, and the Fourier series of an infinite repeating function.

2. It makes trivial work of describing/explaining the properties of *frequency combs*, a very special kind of laser.

Many functions have analytic Fourier transforms. In the following table we show some of the more useful ones.

Function Name	g(t)	$\tilde{g}(\nu)$
Tophat	$\begin{pmatrix} 0, & -\infty < t < -a/2 \\ 1, & -a/2 < t < a/2 \\ 0, & a/2 < t < \infty \end{pmatrix}$	$a\left[\frac{\sin(\pi\nu a)}{\pi\nu a}\right] \equiv a\operatorname{sinc}(\pi\nu a)$
triangle	$\begin{pmatrix} 0, & a/2 < |t| \\ h - \frac{2h|t|}{a}, & -a/2 \le |t| \end{pmatrix}$	$2ah\operatorname{sinc}(2\pi\nu a)$
gaussian	$g(t) \equiv A\exp\left(-\frac{t^2}{\sigma^2}\right)$	$A\sigma\sqrt{\pi}e^{-\pi^2\sigma^2\nu^2}$
constant	C	$\delta(\nu)$
Dirac Delta	$\delta(t - a)$	$\exp(i2\pi\nu a)$
cosine	$\cos(2\pi at)$	$\frac{1}{2}\left[\delta(f - A) + \delta(f + A)\right]$
sine	$\sin(2\pi at)$	$\frac{1}{2i}\left[\delta(f - A) - \delta(f + A)\right]$
Heaviside ($H(t)$)	$\begin{pmatrix} 1, & t \ge 0 \\ 0, & t < 0 \end{pmatrix}$	$\frac{1}{2\pi i f} + \frac{\delta(f)}{2}$
right-side exp. decay	$e^{-|t|/\sigma}H(t)$	$\frac{\sigma^2}{1+(2\pi\sigma\nu)^2}e^{i\phi}, \phi \equiv \tan^{-1}\left(\frac{1}{2\pi\sigma\nu}\right)$
left-side exp. decay	$e^{|t|/\sigma}H(-t)$	$\frac{\sigma^2}{1+(2\pi\sigma\nu)^2}e^{i\phi}, \phi \equiv \tan^{-1}\left(\frac{-1}{2\pi\sigma\nu}\right)$
2-sided exp. decay	$e^{-|t/\sigma|}$	$\frac{\sigma^2}{1+(2\pi\sigma\nu)^2}$

5.3 PROPERTIES OF FOURIER TRANSFORMS

If the Fourier transform could do nothing else but tell us the frequency content of a signal, it would be a very powerful tool. But it can do far more! First a bit of notation. In what follows we assume the following Fourier pairs.

5.3.1 Theorems

Suppose that we define the following Fourier pairs:

$$F_1(x) \rightleftharpoons \Phi_1(p)$$

$$F_2(x) \rightleftharpoons \Phi_2(p)$$

Then the *addition theorem* says that

$$F_1(x) + F_2(x) = \Phi_1(p) + \Phi_2(p).$$

Furthermore, the *shift theorem* says that

$$F_1(x + a) \rightleftharpoons \Phi(p)e^{2\pi ipa},$$

$$F_1(x - a) \rightleftharpoons \Phi(p)e^{-2\pi ipa}, \qquad (5.7)$$

$$F_1(x + a) + F_1(x - a) \rightleftharpoons 2\Phi(p)\cos(2\pi pa).$$

In particular, if $F_1(x)$ is a δ-function, then

$$\delta(x + a) \rightleftharpoons e^{-2\pi ipa},$$

$$\delta(x - a) \rightleftharpoons e^{2\pi ipa},$$

$$\delta(x + a) + \delta(x - a) \rightleftharpoons 2\cos(2\pi pa).$$

This last result should look familiar: the (inverse) Fourier transform of a sinusoidal is a pair of δ-functions, placed symmetrically about the midpoint of the transform. The symmetry of Eqs. 5.1 and 5.2 makes this true, whether or not this is a Fourier transform or an inverse Fourier transform. We just saw this in the above code.

5.3.2 Convolutions

By convolution we mean the integral of the product of two functions. For example, any measuring device will have finite resolution. Perhaps you are using a spectrometer to measure the wavelengths of light coming from a sodium street lamp; even if the theoretical spectrum of the sodium transition lines were effectively δ-functions, your real-life spectrometer would not give you peaks that are infinitely high and infinitely narrow! Your δ-function spectrum would be *convolved* with the *resolution function* of the spectrometer. As a concrete example, suppose the resolution function of the spectrometer were a gaussian of width σ:

$$R(\lambda_0 - \lambda) = \frac{1}{\sigma\sqrt{\pi}}\exp\left[-\frac{(\lambda - \lambda_0)^2}{\sigma^2}\right].$$

Then the measured spectrum, $S_m(\lambda)$ would be the convolution of the "actual" spectrum, $S_a(\lambda)$ with the instrument resolution $R(\lambda)$:

$$S_m(\lambda) = \int_{-\infty}^{\infty} S_a(\lambda')R(\lambda - \lambda')d\lambda'.$$

This can be written more compactly as

$$S_m(\lambda) = S_a(\lambda) * R(\lambda).$$

Convolutions follow many of the familiar rules of algebra:

- The commutative rule: $C(x) = F_1(x) * F_2(x) = F_2(x) * F_1(x)$.
- The distributive rule: $F_1(x) * [F_2(x) + F_3(x)] = F_1(x) * F_2(x) + F_1(x) * F_3(x)$.
- The associative rule: $F_1(x) * [F_2(x) * F_3(x)] = [F_1(x) * F_2(x)] * F_3(x) = F_1(x) * F_2(x) * F_3(x)$

Now here's the amazing part:

$$F_1(x) * F_2(x) \rightleftharpoons \Phi_1(p) \cdot \Phi_2(p) \ , \tag{5.8}$$

where the "$*$" means convolution and the "$\cdot$" means ordinary multiplication. Equation 5.8 means that the convolution of two functions is just the inverse Fourier transform of the product of their Fourier transforms. That is, using our earlier example of spectrometry,

$$S_m(\lambda) = \mathrm{FT}^{-1} \left\{ \mathrm{FT} \left[S_a(\lambda) \right] \cdot \mathrm{FT} \left[R(\lambda) \right] \right\} \ ,$$

where "FT" and "FT^{-1}" mean Fourier transform and inverse Fourier transform, respectively.

Example: Convolution of a Gaussian with a Gaussian

Suppose we define the gaussian function as

$$G(t) \equiv A \exp \left[-\frac{t^2}{\sigma^2} \right]$$

as in Eq. 5.5. Then the Fourier transform of a gaussian is given by Eq. 5.6 to be

$$\tilde{g}(\nu) = A\sigma\sqrt{\pi} e^{-\pi^2 \sigma^2 \nu^2} \ .$$

And the convolution of two gaussians having different widths and amplitudes is

$$\begin{aligned}
G'(t) &= \mathrm{FT}^{-1} \left[g_1(\nu) g_2(\nu) \right] \\
&= \mathrm{FT}^{-1} \left[A_1 A_2 \sigma_1 \sigma_2 \exp \left\{ -\pi^2 \nu^2 \left(\sigma_1 + \sigma_2 \right) \right\} \right] \\
&= \frac{A_1 A_2 \sigma_1 \sigma_2 \sqrt{\pi}}{\sqrt{\sigma_1^2 + \sigma_2^2}} e^{-t^2/(\sigma_1^2 + \sigma_2^2)} .
\end{aligned}$$

That is, you get a gaussian again, but with a width equal to the widths of the two convoluted gaussians, "added in quadrature". This is an extremely useful result. This concludes our discussion of Fourier transforms.

## CHAPTER 6

Special Functions and Other Math Tricks

In this chapter, we discuss some of the important classes of polynomials, which we lump together under the heading "Special Functions". First, though, we will briefly look at an important example of the *Taylor Expansion* which, if you have not already studied it, you will at some point in the physics curriculum. This special case is referred to as a *binomial expansion* and is very important. We will use binomial expansions in our discussion of the special functions that we will cover in this chapter. Note that for all of the math we discuss in this chapter, we will focus on "how to use" rather than derivations.

6.1 BINOMIAL EXPANSION

Consider the expression

$$y = (1 + z)^m . \tag{6.1}$$

Suppose we would like to expand this for a particular value of m. We can do a *Taylor Series Expansion*, giving us:

$$(1 + z)^m = 1 + mz + \frac{m(m-1)}{2!}z^2 + ... + R_n, \tag{6.2}$$

where

$$R_n \leq \frac{x^n}{n!} \times m(m-1)...(m-n+1) \tag{6.3}$$

For example, for $m = 4$, Eq. 6.1- 6.3 would give us

$$(1 + z)^4 = 1 + 4z + 6z^2 + 4z^3 + z^4 . \tag{6.4}$$

The Taylor expansion of a sum of two terms raised to a power is referred to as a *binomial expansion*.

This is an interesting result and plays an important role in probability theory, among other things. But the results are even more interesting if m is a non-integer. For example, suppose $m = -1/2$. Eq. 6.1 would then give us

$$(1+z)^{-1/2} = 1 - \frac{1}{2}z + \frac{3}{8}z^2 - \frac{5}{16}z^3 + \frac{35}{128}z^4 - \frac{63}{256}z^5$$
$$+ \frac{231}{1024}z^6 - \frac{429}{2048}z^7 + \frac{6435}{32768}z^8 - \dots \tag{6.5}$$

We will make use of Eq. 6.5 in the following section.

6.2 LEGENDRE POLYNOMIALS

We will introduce Legendre polynomials using the electrostatic potential of a single coulomb charge. It would be simplest to place the charge at the zero of a coordinate system, but one cannot always do this. For example, imagine if you have multiple charges: unless they are all located in the same place, at most one of them can be at the center of your chosen coordinate system. With this in mind, consider a single charge, as shown in the figure:

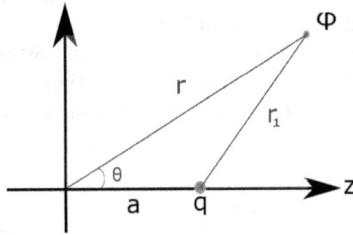

FIGURE 6.1: Showing the geometry of a single coulomb charge, q, and the location of the computed potential, ϕ.

As you know (or will soon learn!) the electro-static potential (ϕ), expressed in mks units, located a distance d from a single charge is:

$$\phi = \frac{1}{4\pi\epsilon_0}\frac{1}{d} \tag{6.6}$$

where ϵ_0 is a physical constant. Now, applying the Law of Cosines to the system in Fig. 6.1, we obtain

$$r_1^2 = a^2 + r^2 - 2ar\cos(\theta) , \tag{6.7}$$

or

$$r_1^{-1} = \frac{1}{r}\left[1 + \left(\frac{a}{r}\right)^2 - \frac{2a}{r}\cos(\theta)\right]^{-1/2} \tag{6.8}$$

Letting $t \equiv a/r$ and $x \equiv \cos(\theta)$, we get, after expanding and collecting terms having the same powers of t,

$$
\begin{aligned}
rr_1^{-1} = &\, 1 + xt + \frac{3x^2 - 1}{2}t^2 + \frac{5x^3 - 3x}{2}t^3 + \frac{35x^4 - 30x^2 + 3}{8}t^4 \\
&+ \frac{63x^5 - 70x^3 + 15x}{8}t^5 + \frac{231x^6 - 315x^4 + 105x^2 - 5}{16}t^6 \\
&+ \frac{429x^7 - 693x^5 + 315x^3 - 35x}{16}t^7 \\
&+ \frac{6435x^8 - 12012x^6 + 6930x^4 - 1260x^2 + 35}{128}t^8 + \dots
\end{aligned}
\tag{6.9}
$$

This can be written more compactly as

$$
rr_1^{-1} = \sum_{n=0}^{\infty} P_n(x)t^n \,,
\tag{6.10}
$$

where the $P_n(x)$ are the coefficients from Eq. 6.9 and are called Legendre Polynomials. Then, comparing Eqs 6.9 and 6.10, we see that the first several Legendre polynomials are given in Table 6.1.

TABLE 6.1: The first several Legendre polynomials and their functional form.

$P_n(x)$	Polynomial
$P_0(x)$	1
$P_1(x)$	x
$P_2(x)$	$\frac{(3x^2-1)}{2}$
$P_3(x)$	$\frac{(5x^3-3x)}{2}$
$P_4(x)$	$\frac{(35x^4-30x^2+3)}{8}$
$P_4(x)$	$\frac{(35x^4-30x^2+3)}{8}$
$P_5(x)$	$\frac{(63x^5-70x^3+15x)}{8}$
$P_6(x)$	$\frac{(231x^6-315x^4+105x^2-5)}{16}$
$P_7(x)$	$\frac{(429x^7-693x^5+315x^3-35x)}{16}$
$P_8(x)$	$\frac{(6435x^8-12012x^6+6930x^4-1260x^2+35)}{128}$

It seems like a lot of work to generate tables of Legendre polynomials – and it was, back in the day. But nowadays, one can use packages like those included in Python's sympy package. Again, a discussion of sympy is beyond the limits of this text, but the command for producing the functional form of the Legendre polynomials is simple enough to demonstrate here. For example, to generate the factors of x in Table 6.1 for a desired m, simply use this 2-line program:

```
1  '''
2  Computing the Functional Form of P_n(x)
3  '''
4  from scipy.special import legendre
5  print('Functional form for P_3(x) = ', legendre(3))
```

Listing 6.1: Sample code showing how to compute the functional form of $P_n(x)$.

When I execute this code, I get:
Functional form for P_3(x) =
$2.5x^3 - 1.5x$

which agrees with $P_3(x)$ in Table 6.1.

Most of the time we do not need the Legendre polynomial expressions; usually we just need the numerical values of Legendre polynomials, given n and x. We therefore show some Python code that computes values of $P_n(x)$ for n in the range of 0 to 5, and x from -1 to +1.

```
1  '''
2  Plotting Legendre Polynomials
3  '''
4  import numpy as np
5  import matplotlib.pyplot as plt
6
7  x = np.linspace(-1, 1, 100)   # Create a range of x values
8
9  # Generate Legendre polynomials up to degree 5
10 for n in range(6):
11     coeffs = np.zeros(n+1)
12     coeffs[n] = 1
13     p = np.polynomial.legendre.Legendre(coeffs)
14     plt.plot(x, p(x), label=f'P_{n}(x)')
15
16 plt.legend()
17 plt.title('Legendre Polynomials')
18 plt.show()
```

Listing 6.2: Sample code showing how to compute values for Legendre polynomials.

When I execute this code, I get:
Technically, Legendre polynomials are the solution to a specific differential equation:

$$\left(1 - x^2\right)\ddot{y} - 2x\dot{y} + n(n+1)y = 0 \ , \tag{6.11}$$

where y represents the function, n is an integer representing the degree of the polynomial, and the dots indicate derivatives of y with respect to x.

You can verify this by directly substituting an expression for P_n, from Table 6.1 into Eq. 6.11, for selected values of n.

Homework 6.1: Verifying Legendre

Verify, by differentiation and direct substitution, that the functional representation of $P_3(x)$ from Table 6.1 satisfies Eq. 6.11.

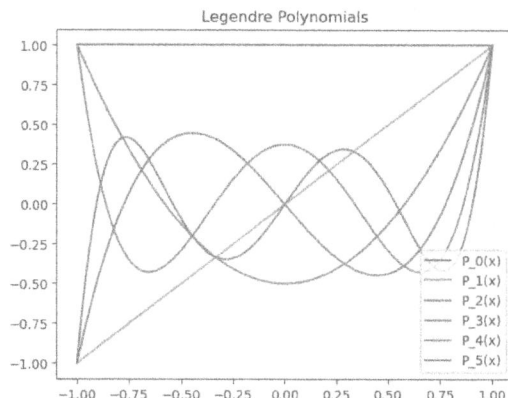

FIGURE 6.2: Plots of the first few Legendre polynomials.

As we have seen, Legendre polynomials are important in determining electro-static potentials due to one or more charged particles. Notice, however, that Fig. 6.1, which we used in deriving expressions for the Legendre polynomials, is symmetric about the z-axis. That is, if we call the angle about the z-axis ϕ (not to be confused with the electrostatic quantity having the same symbol!) we see that the potential is independent of that angle. But suppose that we have a system which is *not* az-imuthally symmetric? This corresponds to a more general case than we have been considering. Instead of a differential equation that has a solution dependent only on one angle (θ), we need an equation whose solution depends upon two angles, θ and ϕ. (Not the potential, the coordinate!) Indeed, if we wish to model a 3-dimensional problem in, say, spherical-polar coordinates, we must have a solution that depends upon *three* coordinates: r, θ, and ϕ.

Such a differential equation arises in quantum mechanics. For example, the Schrödinger's equation for the hydrogen atom is a 3-dimensional partial differen-tial equation. Using a method known as separation of variables, one can express solutions to this equation as the product of two functions, one depending only on r, and the other depending on θ and ϕ. We will now explore this so-called hydrogen wave-equation, with the idea of demonstrating how to break up a partial differential equation into several ordinary differential equations. To do this, we will introduce a technique known as *separation of variables*.

6.2.1   Separation of Variables

Even an introduction to quantum mechanics is beyond the scope of this book. Thus, we will simply introduce an application of *Schrödinger's equation* to the hydrogen atom. We will then use this to demonstrate a technique known as separation of variables.

Learning all the myriad methods one must employ to solve ordinary differential equations is the work of an entire course, and so we will not broach that topic. Rather, we will simply assert solutions to some differential equations and leave it to the reader to verify that the proposed do, indeed, satisfy the equations.

Our starting point is Schrödinger's equation acting on the centrally symmetric electrostatic potential of the hydrogen atom. It is

$$\frac{-\hbar^2}{2m_e} \left[\frac{1}{r^2} \frac{\partial}{\partial r} \left(r^2 \frac{\partial}{\partial r} \right) + \frac{1}{r^2 \sin\theta} \frac{\partial}{\partial\theta} \left(\sin\theta \frac{\partial}{\partial\theta} \right) + \frac{1}{r^2 \sin^2\theta} \frac{\partial^2}{\partial\phi^2} \right] \psi + V(r)\psi = E\psi \ ,$$

(6.12)

where m_e is the mass of an electron, $V(r)$ is the electrostatic potential of the nucleus, and E is the total energy of the system. The symbol $\hbar$ is a physical constant, and the symbol ∂ means a *partial derivative*, that is, the derivative of a function that has many variables. The partial derivative treats as constants all variables *except* the one indicated, which is differentiated in the usual way. In Eq. 6.12, ψ is the hydrogen wave function that we need to solve for. It is a function of r (the radial coordinate), θ (the polar angle), and ϕ (the azimuthal angle). We will assume that

$$\psi(r,\theta,\phi) = R(r)Y(\theta,\phi) \ .$$

(6.13)

How do we know that ψ can be expressed as the product of a function that depends only on r and a function that depends only on θ and ϕ? We don't. In general, you try to use such a product function, and see if it works. If it doesn't you'll probably have no recourse but to solve the partial differential equation numerically. Here, we will find that this product assumption does work.

Inserting this product function into Eq. 6.12 and rearranging gives us

$$Y(\theta,\phi)\frac{1}{r^2} \frac{d}{dr} \left(r^2 \frac{dR(r)}{dr} \right) + R(r)\frac{1}{r^2 \sin\theta} \frac{\partial}{\partial\theta} \left(\sin\theta \frac{\partial Y(\theta,\phi)}{\partial\theta} \right)$$

$$+ R(r)\frac{1}{r^2 \sin^2\theta} \frac{\partial^2}{\partial\phi^2} - \frac{2m_e}{\hbar^2} [V(r) - E] R(r)Y(\theta,\phi) = 0 \ .$$

(6.14)

Notice that the partial derivatives with respect to r are now total derivatives, whereas the partials with respect to θ and ϕ remained. This is because the function that the partials with respect to r are operating on are now functions of a single variable, r.

Multiplying Eq. 6.14 by $\frac{r^2}{RY}$ and rearranging gives us

$$\left\{ \frac{1}{R(r)} \frac{d}{dr} \left(r^2 \frac{dR(r)}{dr} \right) - \frac{2m_e r^2}{\hbar^2} (V(r) - E) \right\}$$

$$+ \left[\frac{1}{Y(\theta,\phi)\sin\theta} \frac{\partial}{\partial\theta} \left(\sin\theta \frac{\partial Y(\theta,\phi)}{\partial\theta} \right) + \frac{1}{Y(\theta,\phi)\sin^2\theta} \frac{\partial^2 Y(\theta,\phi)}{\partial\phi^2} \right] = 0$$

(6.15)

In Eq. 6.15, the quantity within the curly braces depends only on the variable r. The quantity within the square brackets has no dependence on r. Therefore, those two quantities must sum to 0, regardless of the values that r, θ or ϕ take. Thus, if we wish, we can set the curly bracket quantity equal to a constant, k, and the square bracket quantity equal to $-k$, guaranteeing that the sum of those two quantities will be 0. However, mainly because I know which choice of that constant will make our algebra simpler, I will choose that constant to be, not k, but $l(l+1)$. Thus, Eq. 6.15 can be broken up into the two equations

$$\frac{1}{R(r)} \frac{d}{dr} \left(r^2 \frac{dR(r)}{dr} \right) - \frac{2m_e r^2}{\hbar^2} (V(r) - E) = l(l+1) \qquad (6.16a)$$

$$\frac{1}{Y(\theta,\phi)\sin\theta} \frac{\partial}{\partial\theta} \left(\sin\theta \frac{\partial Y(\theta,\phi)}{\partial\theta} \right) + \frac{1}{Y(\theta,\phi)\sin^2\theta} \frac{\partial^2 Y(\theta,\phi)}{\partial\phi^2} = -l(l+1) \quad (6.16b)$$

Equation 6.16a depends only on r and is called the *Radial Equation*. Equation 6.16b had no r dependence but is a function of both θ and ϕ. It is called the *Angular Equation*. The Radial Equation has as its solution the Laguerre polynomials, which we will get to later. However, you can obtain numerical values of Laguerre polynomials using Python, just as we did with Legendre polynomials. We therefore proceed to analyze the Angular Equation.

We start by using the same separation of variables strategy we had started with, namely guessing that

$$Y(\theta.\phi) = f(\theta)g(\phi) \ . \qquad (6.17)$$

Plugging this into Eq. 6.16b, multiplying through by $\sin^2\theta$, and rearranging, gives us

$$\frac{\sin\theta}{f(\theta)} \frac{d}{d\theta} \left(\sin\theta \frac{df(\theta)}{d\theta} \right) + l(l+1)\sin^2\theta + \frac{1}{g(\phi)} \frac{d^2}{d\phi^2} = 0 \qquad (6.18)$$

The first term in Eq. 6.18 only depends on the polar angle, θ, while the second term depends only on the azimuthal angle, ϕ. Thus, each of these terms can be set equal to a constant. Again, I already know what works best, and so I will choose a constant that will be consistent with everything we know about quantum mechanics: I will set the polar angle equation equal to m^2 and the azimuthal angle equation equal to $-m^2$, guaranteeing that the two terms will sum to 0. Shown explicitly, these equations are

$$\frac{d^2 g(\phi)}{d\phi^2} = -m^2 g(\phi) \qquad (6.19a)$$

$$\sin\theta \frac{d}{d\theta} \left(\sin\theta \frac{df(\theta)}{d\theta} \right) + l(l+1)\sin^2\theta f(\theta) - m^2 f(\theta) = 0 \qquad (6.19b)$$

The resulting azimuthal angle equation is just the "simple harmonic oscillator" equation whose solution can be expressed as

$$g(\phi) \equiv g_m(\phi) = A\cos m\phi + B\sin m\phi = Ce^{im\phi} \ , \qquad (6.20)$$

where A, B and C are arbitrary constants whose values will depend on the boundary conditions of the system under consideration. We have also added the subscript m to the function to explicitly show that its value does depend on m. Returning to the polar equation, we see that we have a non-partial differential equation. Our next task is to hammer it into something we can recognize. We start by massaging the first term:

$$\sin\theta \frac{d}{d\theta} \left(\sin\theta \frac{df(\theta)}{d\theta} \right) = \sin\theta \left(\cos\theta \frac{df(\theta)}{d\theta} + \sin\theta \frac{d^2 f(\theta)}{d\theta^2} \right)$$
$$= \sin^2\theta \frac{d^2 f(\theta)}{d\theta^2} + \sin\theta\cos\theta \frac{df(\theta)}{d\theta} \qquad (6.21)$$

Substituting this back into Eq. 6.19b gives

$$\sin^2\theta \frac{d^2 f(\theta)}{d\theta^2} + \sin\theta \cos\theta \frac{df(\theta)}{d\theta} + l(l+1)\sin^2\theta f(\theta) - m^2 f(\theta) = 0 \qquad (6.22)$$

We will now do something that is, admittedly, a bit non-standard. We recognize that we have a pretty intractable equation and this is partly due to the fact that trig functions are more difficult to deal with than other algebraic expressions. With that in mind, we define

$$x \equiv \cos\theta \ . \qquad (6.23)$$

Then, using the chain rule for differentiation,

$$\begin{aligned} \frac{df(\theta)}{d\theta} &= \frac{df(x)}{dx} \frac{dx}{d\theta} \\ &= \frac{df(x)}{dx}(-\sin\theta) \qquad\qquad (6.24) \\ &= -\sin\theta \frac{df(x)}{dx} \end{aligned}$$

and,

$$\begin{aligned} \frac{d^2 f(\theta)}{d\theta^2} &= \frac{d}{d\theta}\left(-\sin\theta \frac{df(x)}{dx}\right) \\ &= -\cos\theta \frac{df(x)}{dx} - \sin\theta \frac{d}{d\theta}\frac{df(x)}{dx} \\ &= -\cos\theta \frac{df(x)}{dx} - \sin\theta \frac{d}{dx}\frac{dx}{d\theta}\frac{df(x)}{dx} \qquad (6.25) \\ &= -\cos\theta \frac{df(x)}{dx} + \sin^2\theta \frac{d^2 f(x)}{dx^2} \end{aligned}$$

Substituting this into Eq. 6.22 and dividing through by $\sin^2\theta$, we obtain

$$\sin^2\theta \frac{d^2 f(x)}{dx^2} - 2\cos\theta \frac{df(x)}{dx} + l(l+1)f(x) - \frac{m^2}{\sin^2\theta}f(x) = 0 \ . \qquad (6.26)$$

Noting that

$$\cos\theta = x \qquad\qquad (6.27a)$$
$$\sin^2\theta = 1 - \cos\theta = 1 - x^2 \ , \qquad (6.27b)$$

then,

$$\left(1 - x^2\right)\frac{d^2 f(x)}{dx^2} - 2x\frac{df(x)}{dx} + l(l+1)f(x) - \frac{m^2}{1-x^2}f(x) = 0 \qquad (6.28)$$

This is the Associated Legendre equation, which for $m=0$ becomes the Legendre equation. The Associated and "regular" Legendre equations are often written as

$$\left(1 - x^2\right)\ddot{f}(x) - 2x\dot{f}(x) + l(l+1)f(x) - \frac{m^2}{1-x^2}f(x) = 0 \qquad (6.29a)$$
$$\left(1 - x^2\right)\ddot{f}(x) - 2x\dot{f}(x) + l(l+1)f(x) = 0 \ . \qquad (6.29b)$$

First note that Eq. 6.29b is identical to Eq. 6.11. Next bear in mind that yet another way to generate "ordinary" Legendre polynomials is by

$$P_l(x) = \frac{(-1)^l}{2^l l!} \frac{d^l}{dx^l} \left(1 - x^2\right)^l \tag{6.30}$$

And to generate an Associated Legendre polynomial from an "ordinary" Legendre polynomial:

$$P_l^m(x) = (-1)^m \sqrt{(1 - x^2)^m} \frac{d^m}{dx^m} P_l(x) \tag{6.31}$$

6.3 BESSEL FUNCTIONS

Bessel Functions have many uses. Some of these are

- Electromagnetic waves in a cylindrical waveguide
- Heat conduction in a cylindrical object
- Modes of vibration on a drumhead or other thin circular membrane.
- Solutions to the radial Schrödinger equation (in spherical and cylindrical coordinates) for a free particle
- Diffusion problems on a lattice
- Position space representation of the Feynman propagator in quantum field theory

Example: A cylindrical conducting cavity of radius a, containing no charges. To take maximal advantage of the symmetry, we will work in cylindrical coordinates. Then, z is along the axis of the cylinder; ϕ is the azimuthal angle around the z-axis; and ρ is the perpendicular distance from the z-axis. Note that we use ρ instead of r. Because the cylinder is conducting, we have the boundary condition

$$E_z(\rho = a) = 0 \ , \tag{6.32}$$

where E_z is the z-component of the electric field. This is so because there are no electric fields in a conductor. As you will learn in a course on electro-magnetic theory, you can use Helmholz' Equation to solve for E_z:

$$\frac{1}{\rho} \frac{\partial}{\partial \rho} \left(\rho \frac{\partial E_z}{\partial \rho} \right) + \frac{1}{\rho^2} \frac{\partial^2 E_z}{\partial \phi^2} + \frac{\partial^2 E_z}{\partial z^2} + \alpha^2 E_z = 0 \ . \tag{6.33}$$

Again, we have a partial differential equation in three variables. With the hope of converting this into three ordinary differential equations, we again try separation of variables.

$$E_z(\rho, \phi, z) = P(\rho)\Phi(\phi)Z(z) \ . \tag{6.34}$$

Substituting this for E_z in Eq. 6.33 gives

$$\frac{1}{\rho P(\rho)} \frac{d}{d\rho} \left(\rho \frac{dP(\rho)}{d\rho} \right) + \frac{1}{\rho^2 \Phi(\phi)} \frac{d^2\Phi(\phi)}{d\phi^2} + \frac{1}{Z(z)} \frac{d^2 Z(z)}{dz^2} + \alpha^2 = 0 \tag{6.35}$$

Taken together, the first two terms are functions of ρ and ϕ only, and the third term is a function of z only. We can therefore set the third term equal to a constant, $-k^2$, and the sum of the remaining terms equal to k^2. This leads to the two equations

$$\frac{1}{z} \frac{d^2 Z(z)}{dz^2} = -k^2 \tag{6.36a}$$

$$\frac{1}{\rho P(\rho)} \frac{d}{d\rho} \left(\rho \frac{dP(\rho)}{d\rho} \right) + \frac{1}{\rho^2 \Phi(\phi)} \frac{d^2 \Phi(\phi)}{d\phi^2} + \alpha^2 = k^2 \tag{6.36b}$$

Except for containing variables and constants having different names, Eqs. 6.36a and Eq. 6.19a are identical, and therefore have the same solutions:

$$Z(z) \equiv Z_k(z) = A \cos k\phi + B \sin k\phi = C e^{ikz} . \tag{6.37}$$

This solution tells us that along the z-axis of a resonant cylindrical conducting cavity, the z-component of the electric field varies sinusoidally with z.

Returning to Eq. 6.36b, we make the substitution $\gamma^2 = \alpha^2 - k^2$. Substituting this into Eq. 6.36b, and multiplying through by ρ^2, we obtain

$$\frac{\rho}{P(\rho)} \frac{d}{d\rho} \left(\rho \frac{dP(\rho)}{d\rho} \right) + \frac{1}{\Phi(\phi)} \frac{d^2 \Phi(\phi)}{d\phi^2} + \gamma^2 \rho^2 = 0 . \tag{6.38}$$

Again we have managed to break the equation into two parts, one depending only on ϕ and the other depending only on ρ. As usual, we set each side equal to the appropriately signed constant, obtaining

$$\frac{1}{\Phi(\phi)} \frac{d^2 \Phi(\phi)}{d\phi^2} = -m^2 \implies \Phi_m(phi) = D e^{\pm im\phi} \tag{6.39a}$$

$$\frac{\rho}{P(\rho)} \frac{d}{d\rho} \left(\rho \frac{dP(\rho)}{d\rho} \right) + \gamma^2 \rho^2 - m^2 = 0 \tag{6.39b}$$

Doing the differentiation in the first term and multiplying through by ρ, we finally obtain

$$\rho^2 \frac{d^2 P(\rho)}{d\rho^2} + \rho \frac{dP(\rho)}{d\rho} + \left(\gamma^2 \rho^2 - m^2 \right) P(\rho) \tag{6.40}$$

We can express this expression in the "dot notation" giving us

$$\rho^2 \ddot{P}(\rho) + \rho \dot{P}(\rho) \left(\gamma^2 \rho^2 - m^2 \right) P(\rho) = 0 . \tag{6.41}$$

Here, m is the order of the Bessel function. There are many versions of Bessel functions. For example, for a Bessel Function of the First Kind, usually represented by $J_m(x)$, m must be an integer and $\gamma = 1$. For Spherical Bessel Functions, m is a half-integer and $\gamma = 1$. The choice of which kind of Bessel function you want is determined by the geometry. Here with a cylindrical geometry, for example, Bessel Functions of the First Kind are the most convenient.

Now, we have not – and will not – go into the many different techniques for solving regular differential equations. But we will mention that some of these techniques involve creating a series solution. One of these, the Frobenius Method, when applied to Eq. 6.41, with $\gamma = 1$, yields

$$J_\alpha(x) = \sum_{m=0}^{\infty} \frac{(-1)^m}{m! \Gamma(m + \alpha + 1)} \left(\frac{x}{2} \right)^{2m+\alpha} . \tag{6.42}$$

Here, $\Gamma()$ means the Gamma function. When the argument of the Gamma function is an integer, $\Gamma(n) = (n-1)!$. But the Gamma function is far more versatile! Its arguments can be pretty much anything, including complex and non-integer. But note that for Bessel Functions of the First Kind, $J_n(x)$, the argument of the Gamma function is guaranteed to be an integer. Given this, we can express the first few of these functions as series expansions. The first few terms of these are given for $m = 0 - 2$ as:

$$J_0(x) = \sum_{m=0}^{\infty} \frac{(-1)^m}{(m!)^2} \left(\frac{x}{2}\right)^{2m}$$

$$= 1 - \left(\frac{x}{2}\right)^2 + \frac{1}{4}\left(\frac{x}{2}\right)^4 - \frac{1}{36}\left(\frac{x}{2}\right)^6 + \dots \tag{6.43a}$$

$$J_1(x) = \sum_{m=0}^{\infty} \frac{(-1)^m}{m!(m+1)!} \left(\frac{x}{2}\right)^{2m+1}$$

$$= \left(\frac{x}{2}\right) - \frac{1}{2}\left(\frac{x}{2}\right)^3 + \frac{1}{12}\left(\frac{x}{2}\right)^5 - \frac{1}{144}\left(\frac{x}{2}\right)^7 + \dots \tag{6.43b}$$

$$J_2(x) = \sum_{m=0}^{\infty} \frac{(-1)^m}{m!(m+2)!} \left(\frac{x}{2}\right)^{2m+2}$$

$$= \frac{1}{2}\left(\frac{x}{2}\right)^2 - \frac{1}{6}\left(\frac{x}{2}\right)^4 - \frac{1}{48}\left(\frac{x}{2}\right)^6 - \frac{1}{720}\left(\frac{x}{2}\right)^8 + \dots \tag{6.43c}$$

But all of this is a great amount of work! Fortunately, we rarely need to use the Bessel's functional form. Instead, we can simply go to Python where we can numerically generate all the different types of Bessel functions. Here is some code that computes, and then plots, Bessel Functions of the First Kind, for orders 0-5. You can judge for yourself how much easier this is than using the functional form of the Bessels!

```
6  '''
7  Computing Bessel Functions of the First Kind: J_n(x)
8  '''
9  import numpy as np
10 import scipy.special as sp
11 import matplotlib.pyplot as plt
12
13 # need this for matplotlib to work in Jupyter:
14 %matplotlib inline
15
16 x = np.linspace(0,15, 500)
17 for v in range(0, 6):
18     plt.plot(x, sp.jv(v, x), label=f'J_{v}(x)')
19
20
21 plt.legend()
22 plt.title("Bessel Functions of the First Kind")
23 plt.show()
```

Listing 6.3: Sample code showing how to compute the functional form of $J_n(x)$.

When I execute this code, I get:

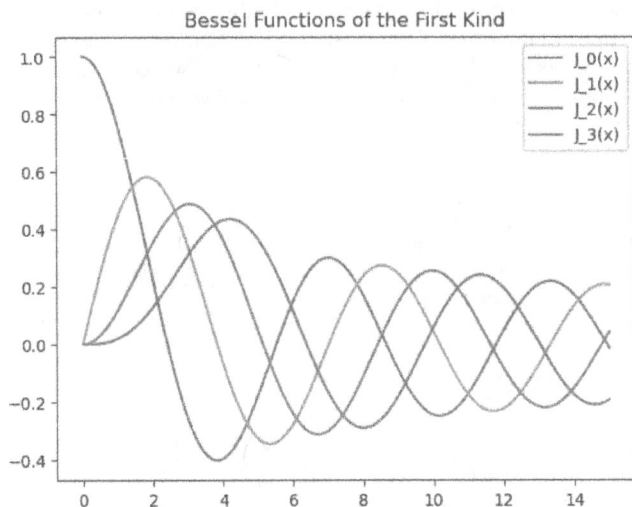

Bessel Functions of the First Kind

FIGURE 6.3: Plots of the first few Legendre polynomials.

You will encounter many other "special functions" in physics, but the ones covered in this chapter seem to be the most pervasive.

Appendices

Installing Python

There are several options for installing Python on your computer, depending on what packages you want and what kind of computer you own. I recommend installing the Anaconda package, which exists for all common computer platforms. While not technically required for you to run Python, Anaconda is a convenient starting point because once you've installed Anaconda, you've installed Python itself, Jupyter, Spyder, and more or less all of the Python modules that you will need for scientific computing.

The first decision you'll need to make is whether to go with an older version of Python or the most up to date. I recommend the latter. As of this writing, the most up to date package is Python 3.x. In this case, you'll want to install Anaconda3. That is, choose the Anaconda version whose number matches the most current Python number.

First point your browser to https://www.anaconda.com/download/. Choose the Linux, Windows or macOS, choose the Anaconda version you've decided on, and hit the install button. Then follow the directions. Easy peasy. If you find at some point that Anaconda does not have some Python package that you'd like to have, the easiest way to install that package is by using the conda package handler from the command prompt. (In Windows, you can open the command window by typing cmd.exe from the start menu; in the macOS you open up Terminal.app; from Linux you open a "terminal" window. In all cases, once you have a command prompt, you simply type

    conda *package*

where *package* is the name of the package you'd like to install. For example, installing Anaconda automatically gives you the powerful graphics package called Matplotlib. But if it didn't, you could load it in by typing

    conda matplotlib

from the command line. It's that simple.

Once Anaconda has been installed, we can use the Anaconda Navigator to launch Jupyter, or Spyder. Jupyter is a sort of "notebook" environment that allows one to make detailed comments, as well as inserting working Python code. When you select Jupyter from the Anaconda Navigator, your default web browser will automatically be invoked and it is from within your browser that you will open, edit, and execute Jupyter modules. I personally prefer doing my preliminary work on a new project from within a Jupyter Notebook because I can heavily document what I am doing and why. Then, once my code is working in its most basic sense, I can always copy and paste it into an *IDE*, or *Integrated Development Environment*. Spyder is the IDE that comes bundled with Anaconda.

Index

For Product Safety Concerns and Information please contact our EU
representative GPSR@taylorandfrancis.com
Taylor & Francis Verlag GmbH, Kaufingerstraße 24, 80331 München, Germany

www.ingramcontent.com/pod-product-compliance
Lightning Source LLC
Chambersburg PA
CBHW060948040426
42445CB00011B/1061